EVERYDAY VITALITY

EVERYDAY VITALITY

Turning Stress into Strength

Samantha Boardman, MD

PENGUIN LIFE

VIKING
An imprint of Penguin Random House LLC
penguinrandomhouse.com

A Penguin Life Book

Grateful acknowledgment is made for permission to reprint the following:

Figure on page 25: Copyright © 2014 by the Robert Wood Johnson Foundation. Used with permission from the Robert Wood Johnson Foundation.

Figure on page 68: Reprinted from *Advances in Experimental Social Psychology*, Vol. 47, page 53. "Chapter One: Positive Emotions Broaden and Build," by Barbara L. Frederickson. Copyright © 2013 Elsevier Inc. All rights reserved. Used with permission from Elsevier.

Image on page 127: © Royal Museums of Fine Arts of Belgium, Brussels / photo: J. Geleyns—Art Photography.

Image on page 187: Reprinted from *Cognition*, Vol. 110, pages 124–129. "Embodied and Disembodied Cognition: Spatial Perspective-taking," by Barbara Tversky and Bridgette Martin Hard. Copyright © 2009, with permission from Elsevier.

LIBRARY OF CONGRESS CATALOGING-IN-PUBLICATION DATA
Names: Boardman, Samantha, author.
Title: Everyday vitality : turning stress into strength / Samantha Boardman, MD.
Description: New York : Penguin Life, [2021] | Includes bibliographical references and index.
Identifiers: LCCN 2021003762 (print) | LCCN 2021003763 (ebook) |
ISBN 9780735222274 (hardcover) | ISBN 9780735222281 (ebook)
Subjects: LCSH: Vitality. | Health. | Stress management. | Well-being.
Classification: LCC RA776.5 .B59 2021 (print) |
LCC RA776.5 (ebook) | DDC 613–dc23
LC record available at https://lccn.loc.gov/2021003762
LC ebook record available at https://lccn.loc.gov/2021003763

Printed in the United States of America
1st Printing

Set in Adobe Caslon Pro
Designed by Cassandra Garruzzo

All names and identifying characteristics have been changed to protect the privacy of the individuals involved.

For Aby, who always sees the whole of the moon.

CONTENTS

PART FOUR: CHALLENGE YOURSELF AND EMBODY VITALITY

PART FIVE: BEYOND YOU

INTRODUCTION

What's wrong?

These two words are a common icebreaker between a doctor and a patient. When someone comes in with pain in the lower right abdomen, an internist will follow this question with a physical exam, take blood, and order a CT scan. If it's appendicitis, surgery will be scheduled to resolve the problem.

Focusing on what's wrong is how doctors formulate a diagnosis and course of treatment. In medical school I was trained to identify markers and manifestations of illness. In anatomy class the professor would identify diseases by calling us over to crowd around a body draped with sheets. One cadaver presented a fatty liver, another cadaver had slim fingers punctuated by oversized, twisted knuckles. The professor pointed out how the fingertips bent downward like a swan's neck—a deformity commonly caused by rheumatoid arthritis. Some of the students referred to this cadaver as "the Swan." But when I leaned in to examine the woman's knuckles, I was struck more by the chipped pink polish on her fingernails. Twenty years later, this detail still tugs at my heart. Our academic training focused on what was abnormal, but this flash of color stopped me cold for the opposite reason: it was *totally* normal.

Psychiatrists also ask, "What's wrong?" but a patient's answers are often vague. There are fewer tests to pinpoint the problem, and it's rare that a simple procedure can fix it. Whatever the diagnosis, most patients who come to see me want the same thing: better days. They long for more connection, joy, and meaning in their daily lives. They want less stress, but they also want more engagement. I tell them that there's no magic wand to erase hassles and banish annoyances. I also warn them to be wary of anyone who insists there is a formula for a stress-free, blissful existence. Negative emotions are part of a full life. Uncertainty cannot be avoided. Stress, frustration, and disappointment are inevitable unless you construct a life so safe and sealed off from reality that you live in a bubble. And the thing about trying to live in a bubble is that bubbles always burst.

Overcoming life's obstacles requires resilience. And what fuels resilience? What allows the bending but not breaking that enables us to spring back? The answer is vitality—the positive feeling of aliveness and energy that lies at the core of well-being.

Vitality is often associated with healthy aging, but it's actually beneficial to everyone. Sometimes defined as "health of spirit," vitality is that sense of feeling psychologically and physically up to a task. Like resilience, vitality is often thought of as a quality people are born with. But possessing vitality is not a matter of luck. Vitality isn't in your head; it's generated by your deliberate actions. Whether you're age twenty or eighty, vitality helps you get the most out of each day. Vitality is associated with positive health outcomes like productivity, better coping with stress and challenges, greater mental health, and the ability to manage negative emotions. In short, as writer and psychologist Andrew Solomon observed, "The opposite of depression is not happiness, but vitality."

This book offers daily strategies for cultivating vitality and making better days. You will learn about the three main wellsprings of everyday vitality: meaningfully connecting with others, engaging in experiences that challenge you, and contributing to something beyond yourself.

Since the pandemic struck in early 2020, grief and loss have been overwhelming—loss not only of loved ones but also of our daily routines, including jobs, celebrations, schooling, social connections, and so much more. Experts have warned of the mental illness tsunami that could follow the medical threat. COVID-19 has also laid bare the disparities that remain embedded in the fabric of America. We need strength and energy to activate long-overdue social change.

Even before the colliding crises of 2020, many people felt like bystanders in their own lives, fulfilling others' demands and suppressing their own desires while meaningful moments passed unnoticed, beauty went unseen, and connections were dropped. Too often our days seem to be both bursting-at-the-seams yet unfulfilling. They have become a thankless game of Whac-A-Mole, but with no chance of winning even a sorry-looking stuffed toy.

"Everything I do these days is a 'have to,' not a 'want to,'" one patient told me.

Another patient explained, "I always ask my husband how his day was and then barely listen to the response. I go through the motions of being thoughtful, but my mind is somewhere else. I'm thinking of an email I have to return or an errand I forgot to run. Yesterday I asked my husband how his day was, and he told me he already answered that question. Twice!"

To manage the excessive demands of daily life, people often react by directing attention inward. Self-focused attention can be productive in the short term. We need self-reflection to process experiences so that we can learn from them and move forward. But too much self-reflection can result in self-absorption, which can entrench us in ruminative thoughts. When this happens, instead of an oasis of revitalization, getting locked in our own heads can become counterproductive, isolating us from others and closing us off from opportunities to expand our minds, exercise our bodies, and stretch our souls.

My experience as a psychiatrist has convinced me that self-immersion is *not* the answer to most problems. Vitality doesn't come from disengaging from the world while you "find yourself." Vitality comes from living well *within* the world.

When I trained to become a psychiatrist, the rigors of residency taught me how to diagnose a major depressive episode and how to tell the difference between bipolar disorder and schizoaffective disorder. I honed my focus on the immediate problem of lessening a patient's pain and spent little time on the Big Questions. Instead of asking, "What can make this person's life better?" I did what I'd been taught and turned my attention to the more immediate question, "How many milligrams of an antidepressant should I prescribe?"

After completing my residency, I continued working at the hospital and then opened a private practice in Manhattan. My private patients were less critical, but many were entangled in relationship issues and job stress. Some were battling depression and anxiety. Others were what I call "air quote fine"—managing to stay above the surface but still feeling the pull downward. Using a combination of medicine and therapy, I aimed to lessen their pain and diminish their angst. I considered myself to be an expert in "undoing."

While Claire's life (*all patient names have been changed*) might look enviable to those on the outside, she was feeling numb and unfulfilled. She had three young daughters, a workaholic husband, and an endless list of responsibilities that included serving as "class mom" and shuttling an ungrateful mother-in-law to doctor appointments. Claire's days were long, draining, and repetitive. She had quit her job as a full-time lawyer but continued doing pro bono legal work that she found "interesting but demanding." She felt isolated and cut off from her old friends. To calm her frayed nerves after a long day, Claire would savor two large glasses of red wine and a wheel of brie cheese with crackers, while watching reruns of *Sex and the City* and sneaking a cigarette . . . or two . . . or three. She knew the routine wasn't healthy, but it brought her relief. At her weekly appointment we would devise strategies to help her feel less overwhelmed, less irritable, and less miserable.

As the weeks passed Claire started making progress. She reported that she and her husband were having fewer fights. She was less judgmental about her mother-in-law and less impatient with her kids. Then, at one session, she looked me in the eye and said something that shocked me.

"Dr. Boardman, I hate coming to our weekly sessions," she confessed. "All we do is talk about the bad stuff in my life. I sit in your office and complain for forty-five minutes straight. Even if I am having a good day, coming here makes me think about all the negative things. I'm done."

And she was. That was our last session. It was also a turning point in my life.

Claire's words stung, but she had a point. The American Psychiatric Association defines psychiatry as "the branch of medicine focused on the diagnosis, treatment and prevention of mental, emotional and behavioral disorders." The emphasis is on illness. There is no mention of what can be done to boost well-being for those with a mental illness, or for those who are "fine" but not thriving. I'd been so trained to focus on what was wrong in my patients' lives and doctoring by the book that my efforts were making Claire feel worse.

The experience with Claire made me rethink what it means to be a psychiatrist. I began to recognize that "undoing" problems is not always connected to a patient's feeling mentally strong. Even if an issue like depression is successfully treated, it doesn't necessarily mean that an individual is living a full and engaged life. I was well versed in dialing down misery but knew little about factors that promote well-being or enable a good day. I had a lot more to learn about what really constitutes mental health.

At the age of forty, with a husband and two small kids, I headed back to school. Upending our family's routine was disruptive and scary . . . and mind-expanding. The University of Pennsylvania has a unique program, Applied Positive Psychology, headed by field pioneer Martin Seligman, who had been grappling with similar issues for a long time. As early as the late 1990s he had called for a new way to think about mental health.

In his book *Flourish*, which was published the same year that I joined the program, Dr. Seligman wrote, "As a therapist, once in a while I would help a patient get rid of all his anger and anxiety and sadness. I thought I would then get a happy patient. But I never did. I got an empty patient. And that is because the skills of flourishing—of having positive emotion, meaning, good work, and positive relationships—are something over and above the skills of minimizing suffering."

During my residency I had learned how to diagnose disorders and to prescribe medications to reduce my patients' misery. That year at Penn rewired my thinking. I studied resilience, optimism, and post-traumatic growth. I explored the role of lifestyle and psychosocial factors that impact mental health. I learned about evidence-based interventions that promote well-being. Dr. Seligman introduced me to Dr. Dilip Jeste, a renowned psychiatrist who had been studying happiness in aging patients with schizophrenia. Strikingly, Dr. Jeste and his team found that almost 40 percent of patients with schizophrenia reported being happy all or most of the time. The levels of happiness felt by these patients was associated not with the severity of their disease but with the presence of positive psychological factors such as resilience, social engagement, optimism, and mastery. In patient after patient, I discovered that it's possible to find wellness within illness, happiness within grief, and strength within stress.

After graduation I returned to my work, my brain bursting with new approaches to old problems. I started giving talks to hospital employees about ways to simultaneously manage stress and build strength—not just in their patients but within their own lives. (Health care workers have been called "the most stressed workers" in the nation.) I launched a blog, positiveprescription.com, two months later with the mission to bring science-backed information about positive mental health to those who don't have the time to read scientific journals yet are hungry for actionable and reliable insights from up-to-date research. Feedback posted on the blog revealed how many people were yearning for more meaning, connection, and engagement.

Today I think of myself as a positive psychiatrist, concerned equally with promoting positive mental health and with fixing my patients' problems. In my clinical practice, I value well-being and resilience as much as pathology identification and symptom reduction. I believe vitality is an important component of well-being and is the heart of everyday resilience. I also believe that vitality is a skill that can be learned and practiced.

In many cases my advice runs counter to conventional wisdom. The contemporary emphasis on self-focus flies in the face of research that meaningful connections and other-oriented actions are what fortify us. We are told to live in the moment, seek pleasure, and avoid discomfort of all kinds. What helps us feel moored in the rushing stream of life is when we're learning, growing, and challenging ourselves.

In addition to drawing on the latest research, a lot of my advice is based on my own experiences working with countless patients over the years and listening closely to their stories. My advice also reflects my perspective and experiences as a mother, a wife, a sister, a friend, and a person who has been in therapy since college.

Everyone deserves to answer yes to these statements:

I look forward to each day.
I almost always feel alert and awake.
I have energy and spirit.
I feel alive and full of vitality.

PART ONE

Cultivate Vitality

Chapter 1

THE PEBBLES IN YOUR SHOE

Supposedly, the worst time to get sick is in July, when brand-new interns flood the teaching hospitals. Although these interns have gained hands-on experience in medical school, dealing with patients on wards is a different proposition. These young doctors are expected to make decisions—big and small—all on their own. They decide which antibiotic to prescribe, whether or not to order a CT scan, how to stabilize a decompensating patient, and how to speak to an anxious family.

When I graduated from medical school in late June 2000, the immediate upgrade to "Samantha Boardman, MD" boosted my confidence. After all those years wearing a short white student coat, I was finally allowed to put on a long one and walk proudly down the hospital hall.

My honeymoon as the newly minted Dr. Boardman was short-lived. As soon as I arrived for the first night shift, a nurse paged me. A patient had died, and a doctor was needed to fill out a death certificate, which I'd only done once before. It was also my responsibility to notify family members and explain that they could request an autopsy if they wished. Of course, no one *wants* an autopsy, so I'd be careful to choose my words.

My beeper went off again. A different nurse informed me that I had to draw blood cultures from a leukemia patient who had a fever of 103 degrees.

"I'll be there right away," I said.

When my beeper alerted me again, it was a third nurse, who was concerned about a patient with a rapid heart rate. By now my own heart rate was rising, too, and the list of patients in need of my attention only continued to grow. My initial excitement at having become "Dr. Boardman" evaporated. Imposter syndrome kicked in. I felt like a fraud, playing "dress up" in my white coat. My face turned red, and tears pricked my eyes. I swallowed a big gulp of lukewarm black coffee and took a bite of a doughnut I had shoved into my pocket earlier that day.

If a patient had been in a cardiac crisis, I would have known exactly what to do. I would have swung into action: check the patient's pulse and breathing. If the patient was in distress, I would call a CODE to alert the Rapid Response Team that a patient was in cardiac arrest and then begin chest compressions. I was well trained in the handling of major emergencies, but this was an unrelenting barrage of smaller challenges.

"It isn't the mountains ahead to climb that wear you out, it's the little pebble in your shoe," Muhammed Ali once said.

I could have handled a boulder, but the avalanche of tiny pebbles was overwhelming.

I hear echoes of my first night working as an intern in the stories of many patients. They tell me that they are emotionally exhausted and feel pulled in a thousand different directions at once. Although their schedules are jam-packed, they long for a sense of genuine fulfillment. When I ask them to describe their state of mind in a single word, they reply with "drained," "depleted," and even "dead."

"It's not that I am drowning," a patient in her early forties explained to me. "I am keeping my head above water, but the waves are big and I am getting constantly getting splashed in the face."

Another patient summed it up this way: "Every day feels like a Sunday night—full of dread and emptiness." When I asked him to describe his mood in one word, he thought a moment and replied, "lackluster."

Another patient quoted a line from a Robert Lowell poem to convey her exasperation with the demands of daily life: "How will the heart endure?"

The Pebbles Can Pummel You

Determining whether a person is clinically depressed is not an arbitrary decision. Psychiatrists follow strict guidelines specified by the DSM (Diagnostic and Statistical Manual of Mental Disorders) and look for at least five of the following nine symptoms lasting at least two weeks:

1. Feels depressed most of the day, nearly every day, as indicated by subjective report (e.g., feels sad, empty, hopeless) or observation made by others (e.g., appears tearful)
2. Feels markedly diminished levels of interest or pleasure when engaging in all, or almost all, activities most of the day, nearly every day (as indicated by subjective account or observation)
3. Significant weight loss when not dieting, or weight gain or decrease, or increase in appetite
4. Sleep disturbance
5. Psychomotor agitation or retardation nearly every day (observable by others, not merely subjective feelings of restlessness or being slowed down)
6. Fatigue or loss of energy
7. Diminished ability to think or concentrate, or indecisiveness
8. Feelings of worthlessness or excessive or inappropriate guilt nearly every day
9. Recurrent thoughts of death (not just fear of dying), recurrent suicidal ideation without a specific plan, or a suicide attempt or a specific plan for committing suicide

I include these criteria not only because I want readers to know that depression can manifest in many ways, but also to underscore the importance of seeking professional help if they apply to you or a loved one. Over the years I have diagnosed, hospitalized, and treated many patients with the full range of the symptoms described above. But there are also many who qualify for an "almost diagnosis"—not mentally ill by clinical standards but lacking positive mental health.

When I first opened my private practice, most of the new patients I took on were at an inflection point. They sought help to assess a life-changing decision or to understand a relationship, or they were in the midst of a significant transition, often following a loss. The chronic issues in their daily lives did not take center stage. Today more and more patients come to see me *because* of the ups and downs in their daily lives. They are feeling worn out and worn down by the daily grind.

Women seem to feel it the most. Almost half of the women surveyed said they frequently experience daily stress, and more than 40 percent said they feel as if they don't have enough time. Their lives are nonstop, with a to-do list that seems bottomless. Often a lack of vitality only amplifies their stress. Patients often just give up and sigh, "I guess that's just life."

The hassles of day-to-day living—the annoying, anxiety-provoking, and frustrating experiences that are embedded into everyday life—are a significant source of stress. Seemingly minor occurrences—an argument with a child or partner, an unexpected work deadline, arriving late for an appointment, missing a train, or dealing with a malfunctioning computer—all contribute. One study's results indicated that watching the news and losing your cell phone are among the top ten daily events that stress people out. Even a long line at your local coffee shop or not having hot water for your morning shower can be enough to put you in a terrible mood.

We know it's absurd to allow something minor to ruin a minute, let alone a day. We try to dismiss these daily irritations as irrelevant or as the "first-world problems" they are. We tell ourselves that they don't matter in the long run. But they do.

Many assume that major life events like divorce, the death of a spouse, and the loss of a job are the most virulent causes of stress, but a University of California, Berkeley, study confirmed that so-called microstressors are the ones we need to watch out for: "[T]hese kinds of stressors have been taken for granted and considered to be less important than more dramatic stressors. Clinical and research data indicate that these 'micro-stressors' acting cumulatively, and in the relative absence of compensatory positive experience, can be potent sources of stress."

The impact of challenges that occur during everyday living on both a person's physical and mental health cannot be underestimated and are, in fact, better predictors of health than major life events.

Researcher Richard Lazarus was one of the first to recognize how relatively minor incidences could have so strong an impact. He believed that the overall level of demands on a person and the perception of resources available to meet those demands were what determined if a potential hassle became an actual one. So if an individual is already feeling taxed, an event that might be typically ignored or overlooked—a leaky pen, a missed subway, spilled coffee—takes on a far more negative tone. If the individual is feeling strong, such annoyances may be easily shrugged off or simply pass unnoticed.

During another study, people were asked to record their daily microstressors in a diary, and the results concluded, "Minor daily stresses affect well-being, not only by having separate, immediate, and direct effects on emotional and physical functioning, but also by piling up over a series of days to create persistent irritations, frustrations, and overloads that may result in more serious stress reactions such as anxiety and depression."

Even routine and relatively predictable anxiety-inducing situations can impact health. Watching a stressful soccer match, for example, can more than double the risk of having a heart attack. Our immune system is also vulnerable to stress. Students who were about to take a medical board exam were shown to have a less robust response to the hepatitis vaccine than students who received the shot when they were relaxing during the holidays. People who report a great deal of daily stress tend to be more susceptible to the common cold. When healthy volunteers were exposed to nasal droplets containing the flu virus, those who reported ongoing daily stress were more likely to get sick and display worse symptoms. (To quantify mucus production, the researchers weighed the snot in each tissue. I always felt sorry for the poor graduate students who performed this decidedly unglamorous task in the name of science.)

No one needs a study to prove that stress makes it harder to get along with family members and friends. When emotionally exhausted, we feel more irritable and argumentative. We participate less in volunteer organizations and feel less engaged at work. General well-being beyond social situations takes a hit. People sleep less, eat more, exercise less, have less sex, and watch TV or play video games more than usual. A gravitational pull toward self-interest and other unproductive patterns of thought and behavior kick in. These responses aren't good for the individual nor for those around them.

A Demanding Boss, a Sick Cat

Bella was referred to me by an internist after she burst into tears at his office. She was twenty-nine years old and worked at a clothing company in Manhattan. She lived with her boyfriend, Joe, but they didn't spend much time together. Between their work and travel schedules, she said they were "like passing ships in the night." When they were together, they bickered over household matters. Joe had recently forgotten to pick up cat food on his way home, and a nasty argument ensued. She interpreted his forgetfulness as a sign he was no longer invested in their relationship.

When she came to see me her chief complaint was, "I'm just frazzled." She mostly enjoyed her job and was productive, but she always felt pressed for time. She was eager to find herself, figure out who she was. "All I want," she told me, "is to be happy." But every day was nonstop aggravation. Her commute was a nightmare, because she had moved out to Long Island for the improved "quality of life" and now laughed at the irony. Long work days, a demanding boss, and a sick cat topped a list of obligations and responsibilities. She felt isolated, and her now-stale relationship with Joe made her feel even more alone.

Bella prided herself on not needing much sleep. On weeknights she would watch MSNBC until one o'clock in the morning. During weekends she would crash, staying in bed for most of the day and watching movies. She said she had no time for exercise or to cook. When she was at work, she mostly ate from the vending machine or ordered a cheeseburger from the greasy spoon down the street. She rarely saw her friends and had even canceled attending her college roommate's thirtieth birthday, explaining, "I just didn't have the energy."

One day, she discovered that her driver's license had expired, and she burst into tears at the internist's office. "It had been a rough day to begin with," she explained. "I was late for work. The cat needed to go to the vet . . . and then the license." She was embarrassed at having cried over something "so dumb and minor." She shrugged. "I guess it was just the straw that broke the camel's back."

Bella did not have a history of prior episodes of depression. She was not facing a major crisis. She hadn't been flattened by a big boulder, but the relentless barrage of pebbles had become too much.

When feeling threatened or overwhelmed, we have a tendency to pull away from others and draw into ourselves. Self-focused thinking and behavior served our ancestors well when their survival depended on outrunning a saber-toothed tiger, but these reactions are not necessarily helpful for modern life. Ironically, how people respond to daily stress is often the opposite of what would give them strength. Choices like canceling plans with friends, eating comfort food, staying up late watching television, and skipping the gym offer temporary relief but further deplete vitality.

The key to finding wisdom and strength is not to step away from it all but to learn how to stand firm in the face of it all. For years I have been studying people who overcome daily stress by looking outside of themselves for support, advice, and inspiration. They achieve relief in a variety of ways:

- They override the inclination toward self-defeating action and replace it with an "other-orientation" or "outer orientation," which enables them to consider the thoughts, needs, and experiences of others. They become less defensive about their own choices and more likely to accept and implement helpful advice.
- They don't shy away from discomfort.
- They make plans while remaining flexible.
- When feeling vulnerable, they don't go it alone but reach out.
- They ask questions and keep an open mind.
- They know that their actions affect how they feel.
- They know that their participation is as important as their mindset.
- They work hard at personal relationships, at adding value, and at staying engaged.
- They build a scaffolding of protection and support that fortifies and revitalizes them on an ongoing basis.

Telling Bella, or any patient, what changes to make or how they should conduct their lives undermines their autonomy and rarely leads to the desired result. To paraphrase psychologist Albert Ellis, "shoulding" on ourselves or others is never helpful. Given that it would be impossible to stop the onslaught of all the pebbles in her life, the key to guiding Bella was to help her find the strength and stamina to manage them.

In our sessions, Bella tended to focus on the "noise"—the little annoyances that were bringing her down. She was prone to sharing a lot of "Can you believe it?" anecdotes, such as, "Yesterday, I went to Starbucks to pick up coffee and a scone. After waiting ten minutes, the barista told me that the person right in front of me had just bought the last one. Can you believe it? Why would he tell me that? It pissed me off even more."

Because scone distribution is beyond my control and rehashing the experience only reawakened her irritation, I tried to shift her focus and broaden the discussion. I asked what she hoped for in her life.

Bella said she wanted to feel loved. She wanted to feel accomplished. She wanted to feel physically and emotionally strong. Having established the "what," we then turned to the "how."

There's a cliché that all people ultimately want is something to do, somewhere to live, and someone to love, but what this scenario looks like for every individual is different. And even for the same person, that vision shifts throughout life. As my old therapist used to say whenever I would get too goal-focused—"Remember, Samantha, it's all a process."

Chapter 2

TIRED, STRESSED, BORED

Many of my patients are stuck in "As soon as . . ." lives, spending their days checking off boxes on to-do lists while putting off the lives they would like to lead.

> *"As soon as things calm down, I will make more of an effort with my friends."*
>
> *"As soon things calm down, I will start volunteering."*
>
> *"As soon as things calm down, I will exercise."*
>
> *"As soon as things calm down, I will take that class I am interested in."*

"I feel like I have too many tabs open in my brain," is how Caroline described her typical emotional state. She felt listless and uninspired. At a recent checkup Caroline's internist had told her that she found no evidence of anything physically wrong with her.

"You should get rid of the stress in your life," the internist said, as if that were a realistic possibility.

When I met her, Caroline existed in a holding pattern, circling around fantasies of a different life. "As soon as this project at work is done," she told me, "I am going to rearrange my life." The idea of pressing a magic Eject button to escape from her life was appealing. She admitted she sometimes daydreamed about being diagnosed with an illness that required extended bedrest, which would finally give her a chance to read all the books piled up on her nightstand. She could catch up on sleep. She could have time for real conversations with her friends. She could call her grandmother and really listen to her stories about the good old days instead of responding, "Uh-huh, uh-huh," while simultaneously checking her email.

She felt terribly guilty about wishing an illness on herself. "Only a sick person thinks this way. What's wrong with me? Do I need medication?" she asked. Once again, the absence of a diagnosable problem did not mean she felt vital and strong.

Unfortunately, everyday life has a tendency to conspire against vitality. In a national survey high school students were asked, "How do you currently feel in school?" and the top three answers included "tired," "stressed," and "bored."

When I give talks at hospitals, employees often use the same three words to describe how they feel most days. "Annoyed," "pressed for time," and "disconnected" also come up a lot. All these adjectives are the antithesis of vitality.

When you're feeling sapped of energy, it's natural to look for a quick fix. As an intern, after working all night, my go-to breakfast was a fried egg and cheese on a buttered roll from the corner deli at 5:30 a.m. I believed it was the only way I could get through morning rounds. The irony, of course, is that instead of giving me a boost, the greasy-spoon breakfast was making me feel worse. Within fifteen minutes, the blood-sugar spike caused by the carbohydrates would inevitably crash, inducing lethargy and sluggishness.

Academic journals are filled with studies about the paradox of happiness: why so many people don't do what will make them feel better. A fried egg and cheese sandwich is delicious and promises lasting contentment, but as we all know, the gratification evaporates the instant you take that last bite. Like so many other fleeting indulgences, a greasy-spoon breakfast does not provide an enduring uplift.

Cotton Candy for the Soul

Every single day there are countless scenarios in which we misjudge what will replenish or fortify us. After spending nine hours at your desk and another hour getting home, it's understandable that all you want to do is collapse onto the sofa. A friend calls and invites you out to meet her new boyfriend. You decline because you just don't have the energy. Spending the evening in front of mind-numbing television is so much easier and requires so little.

But as much as doing nothing seems like the perfect way to unwind, the results of a study, "The Guilty Couch Potato," found that people who turned to screens as a strategy to decompress didn't feel better afterward. In fact, instead of becoming relaxed and recovered, they felt even more drained and reported decreased vitality. Patients often tell me how they wind up depleted after spending an evening losing themselves in a video game or scrolling through Instagram on a smartphone.

I describe these devitalizing activities as "cotton candy for the soul." The first bite tastes good, but by the time you have polished off the pink tornado, your tongue hurts and your hands are sticky. You are filled with sugar and regret. On top of feeling gross, you're still hungry.

As a psychiatrist, I consider it an important part of my job to help patients minimize the "empty calories" in their daily lives and instead make choices that nourish them. I ask them to dissect their day to give me a glimpse of how they actually live: What captures their attention? How do they make their decisions and allocate their time? I ask about their hassles and what gets under their skin. I inquire about what brings them joy, satisfaction, and fulfillment. I ask them to walk me through their routines and to tell me about their habits and rituals. I do my best to understand "the doings" of their daily life. Ultimately, I am interested in exploring the following questions with my patients:

- Do their daily actions reflect what matters to them?
- Do they behave intentionally? Or do things "just happen?"
- Are they doing what makes them feel strong?
- Do they make fortifying and vitality-building choices?
- Do they embody what they care about?

I believe that vitality is cultivated and enhanced through productive and meaningful actions: having a good conversation, doing a favor for someone, going for a walk, reading an interesting article and then calling a friend to discuss it. These commonplace experiences and micromoments are the building blocks of everyday resilience. They are other-oriented. They are outward-oriented. They are action-oriented. They are not internal, nor individual, nor do they require sustained self-immersion. On the contrary, they require engagement and interaction.

Vitality involves intersecting with and participating in the world around you. It is not predicated on taking a year off to find yourself. It doesn't require making a drastic change. You don't need to lose yourself in self-reflection. You don't need to overhaul your existence, or reinvent your life, or wait until the chaos surrounding you settles down. These ordinary, simple shifts in behavior can easily become a part of your daily routine.

Those who know how to handle the ups and downs of daily life with agility and grace are masters at overriding the carapace of self-immersion. They are adept at overcoming the temptation to retreat into themselves and become submerged in their innermost thoughts. Often this means doing the opposite of what they feel like doing. They get off the couch. They meet their friends. They put down their phones. They do so by putting their values front and center, making deliberate choices, and structuring their life so as to choose healthier, intentional options.

When I meet new patients, I ask them to list the three things they value the most in their lives. Many say, "I value being a good parent . . . a good partner . . . a good sibling . . . a good son or daughter . . . a good friend . . ." Many tell me they value their health, volunteering, learning new things, and being a good person. Next I ask them to fill in a pie chart of what they actually *do* and how they spend the hours of their day. They are often surprised to discover how many hours, even free time, are spent returning emails, surfing the web, updating Facebook, and checking Instagram, even though those activities were nowhere near the top of their priorities list.

The idea behind this exercise is to encourage more overlap between what they care about and what they actually do. I have noticed that the more patients walk their walk the less they are bothered by pebbles. And many gain a spring in their step.

Encouraging people to think about what they genuinely value redirects their gaze and diverts self-focus. Research shows that an exercise of affirming one's values generates lasting benefits in creating positive social feelings *and* behaviors. People who put their values front and center aren't just more attuned to the needs of others, they truly make better connections. Plus, making values a more explicit part of life increases problem-solving skills and helps manage stress. Struggling students who wrote briefly about their values at the beginning of a semester got better grades at its end. Taking a moment to think about something that is important to you is a strategy you can roll out before entering any high-pressure performance situation.

Insight Imperialism

What this exercise really brings home is how important our everyday actions are. It is often assumed that thoughts govern our existence and, therefore, well-being begins in the mind. People come to therapy to increase self-awareness and better understand what is going on in their heads. They hope for insight and self-discovery. Margaret Thatcher's famous quotation captures this idea that one's thoughts are all powerful.

> *Watch your thoughts for they become words.*
> *Watch your words for they become actions.*
> *Watch your actions for they become habits.*
> *Watch your habits for they become your character.*
> *And watch your character for it becomes your destiny.*
> *What we think, we become.*

This mindset informs most therapeutic processes as well. Therapists work with patients to gain greater understanding of who they are and what they want. Through interpretation and reflection, patients begin to initiate the change they long for in their lives. But an overriding emphasis on insight at the exclusion of everything else, dubbed "insight imperialism" by Paul Wachtel, misses the essential role of everyday actions and patterns of living.

Thinking and talking about issues can only take you so far. You can reflect ad nauseam on your internal world, but it's your actions and experiences in the real world that shape you.

We are, in fact, what we do.

Chapter 3

LITTLE *r* RESILIENCE

Two recent conversations made me realize that the word *resilience* has reached cultural saturation. First, my hairdresser told me to try a new "resilience-enhancing shampoo," and then my vet suggested that my dog Panda needed to be more resilient when she encounters other dogs. (She is rather skittish.)

But when I was in medical school, studying to be a psychiatrist, the concept of resilience was barely mentioned. Back then resilience was not considered to be an ability that could be learned or cultivated. The general belief was that anyone coping well in the wake of a trauma or personal loss was either extraordinary or in denial. If a widow didn't express "sufficient" grief, she was thought to be either coldhearted or avoiding the grim reality of the loss. In our culture the "merry widow" has long been the subject of gossip and suspicion. Detectives in mystery novels have taught us to monitor a widow for any hint of good cheer lest she prove to be the murderer!

During my training I was taught that anyone who endured a major life event would require some form of clinical intervention, whether medication or talk, to manage their distress, but research has shown that this isn't the case at all. Psychologist George A. Bonanno found that other than some short-lived disruptions, most people navigate their way through difficult experiences with minimal impact on their functioning. They are able to go to work. Their relationships don't suffer. They remain capable of experiencing positive emotions like gratitude and love. And they are able to move forward with their lives and take on new challenges. In short, many do just fine.

Bonanno's extensive research includes individuals who experienced the 9/11 terrorist attack, Hong Kong residents affected by the SARS epidemic, residents living with chronic stress in the Palestinian territories, and people who have lost their life partner. Remarkably, of all reactions to trauma and loss, resilience is the most common. It is more ordinary than extraordinary.

I call this Big *R* resilience—the kind that's defined as "the process of adapting well in the face of adversity, trauma, tragedy, threats, or significant sources of stress."

Studies of former prisoners of war and civilian survivors of severe psychological trauma have helped researchers identify some of the factors that help people persevere. These individual resilience factors include

- the capacity to make realistic plans and carry them out;
- a positive view of oneself;
- confidence in one's strength and abilities;
- communication and problem-solving skills; and
- the capacity to manage strong feelings and impulses.

Big *R* resilience is about bouncing back from major traumatic events that are relatively rare. What psychiatrists need to pay more attention to is "little *r* resilience." This is the resilience that enables us to deal with the daily grind. As Anton Chekhov observed, "Any idiot can handle a crisis—it's the day-to-day living that wears you out." A study of more than twelve hundred people concluded that those who reported a high number of everyday hassles were three times more likely to die than those who reported lower levels. So much for the old Nietzsche maxim: what doesn't kill you makes you stronger.

People: Velcro versus Teflon

What bothers one person might not bother another. Having to walk the dog might be perceived as an annoyance or a pleasure, depending on the weather, the time, the walker's feelings about the dog, not to mention whether or not he has issues handling a pooper scooper. Like beauty, bother is in the eye of the beholder.

There are also differences in how individuals respond to irritations. Some are more easily rattled or "reactive," a term psychologists and psychiatrists use to describe individuals who respond with greater negativity. When things don't go their way, reactive people are more likely to blame themselves, to become more distressed and withdrawn, and to feel more overwhelmed. They also tend to have a harder time letting go of their negative emotions. The residue of a bothersome experience—an argument with a colleague or getting stuck in a traffic jam—lingers past the incident and can ruin a day . . . or even more.

Professor David Almeida divides people into two categories: Velcro and Teflon. "With Velcro people, when a stressor happens it sticks to them; they get really upset and, by the end of the day, they are still grumpy and fuming," he explains. "With Teflon people, when stressors happen to them, they slide right off. It's the Velcro people who end up suffering health consequences down the road." In a ten-year study, Professor Almeida and his team at Penn State University found that people who are easily upset in the moment and continue to dwell on their negative feelings are more likely to experience subsequent health problems such as pain, arthritis, cardiovascular complications, and mental health issues.

A friend who studies the history of science once told me that seventeenth-century scientists believed that when people became upset, their blood literally boiled. We now know that our blood isn't actually cooking, but physiological changes do occur when you encounter a stressor. Your heart rate increases, blood pressure rises, and cortisol is released. If whatever upset you remains at the forefront of your thoughts, your body maintains this stress response.

Like Velcro, stress can stick to you. Plus, a bad mood affects more than just your state of mind. Negativity impacts behavior and can result in less physical activity, poor food choices, social withdrawal, avoidance, and disrupted sleep. And the fallout of blood boiling extends beyond the moment. It may sound absurd, but how you react to getting a parking ticket today can predict your health a decade from now.

Why are some people's feathers more easily ruffled than others? Temperament certainly plays a role. People with highly sensitive emotional systems react with more frequent and intense negative emotions and experience them for longer. Feeling depleted, devitalized, and emotionally exhausted also amplifies the stress response.

We are all familiar with the makings of a bad day—disruptions, hassles, conflicts, negative thoughts, and bad news are at the top of a long, long list. But what are the factors that lift our spirits and keep us going? Researchers at the University of Rochester studied these questions and discovered three answers. (Incredibly, chocolate didn't factor into them.) Building vitality relies on boosting the following three basic psychological needs:

1. **Autonomy** is the experience that you are the author of your own behavior and free to make your own choices. It is the opposite of feeling controlled or like a pawn. Autonomy protects against feelings of frustration and dissatisfaction. It involves being proactive, planning ahead, and making decisions that reflect your values.
2. **Competence** is the experience of feeling effective at what you do. It derives from activities as big as completing a meaningful project to actions as simple as making the bed or working on a hobby. Competence protects against feelings of hopelessness.
3. **Relatedness** is the experience of feeling close and connected to others. The more people engage in meaningful conversation and

the more they feel understood and appreciated on a given day, the more related they feel to their partners or friends.

According to self-determination theory, first introduced by Ed Deci and Richard Ryan, these three fundamental needs are essential for human growth, integrity, and health. The satisfaction of these needs sheds light on the not particularly surprising finding that people are happier during weekends, which present more opportunities for rewarding social interaction (relatedness) and fewer obligatory and more self-directed activities (autonomy) than during the workweek. (Competence seems to remain relatively stable throughout the weekly cycle.) If during the week you engage deliberately in activities that increase your autonomy and forge more satisfying social connections, you will usually experience less of a "weekend effect." You will also be inclined to feel less dread about Mondays.

When I first met Gina, she reminded me of Eeyore in *Winnie the Pooh*—all gloom and doom. She told me that she came by it naturally. She was just like her mother—always nervous, always worried, and extremely sensitive to criticism and rejection. If there were Velcro levels, she was industrial strength. Gina took an online personality test that determined she was "extremely neurotic." She didn't question that conclusion, but assumed it was her destiny. Everything supported this view, including a magazine article that explained people have an emotional set point and that well-being is genetically determined. "I read that if I win the lottery or become paralyzed, after a few months, I will return to the exact same level of unhappiness," she recalled.

Gina had come to me because she was having problems sleeping and not because she was interested in or believed in change. Other doctors had advised her to meditate and do yoga. She tried, but claimed meditation made her more anxious and yoga hurt her back. One doctor suggested removing the television from her bedroom, which Gina deemed "insane." Sound machines annoyed her.

I took a different approach. Instead of focusing exclusively on what I thought Gina could do to reduce stress, I asked her to think of activities that might help her feel more in control, more connected, and more competent. The key was that these actions be meaningful to her and self-generated.

We discussed various possibilities, and she decided to take charge of her sleep by initiating a nighttime ritual, beginning with a "power down" hour. At 9 p.m. she would place the remote control for the bedroom TV and her cell phone in the kitchen. Gina also decided to stop drinking Diet Coke or coffee after noon. It was her idea to join a morning jogging group, and she set a goal of running a six-minute mile.

Gina's nights soon got better, and her days did, too—because she was not only getting more sleep but also engaging in activities that were satisfying her needs for autonomy, competence, and relatedness. Her mood was brighter, and she was less reactive to the irritations she encountered. Although the roll of the genetic dice may have endowed her with a Velcro temperament, she was becoming more Teflon by the day. Our genes certainly play a role in shaping us but, as Gina's case demonstrates, we are more than just our genes. How we spend our time impacts who we are.

Reducing Mountains to Molehills

Focusing on a patient's innermost thoughts and feelings along with medication management was my go-to treatment strategy for many years. I encouraged patients to reflect so that they could better understand themselves and their inner conflicts. When I treated someone who was depressed, anxious, or simply stressed out, I assumed that the source of the problem resided in a patient's brain or mind.

For those who were having trouble managing the daily grind, I focused on stress reduction. We spoke at length about what was bothering them to help them minimize conflict and to lessen aggravation. On an as-needed basis, I prescribed Ambien for trouble sleeping and Effexor, with a little Klonopin, to dial down panic or anxiety. I prescribed Wellbutrin to target symptoms of depression.

My childhood caregiver, whom I adored, had a standard response that I can still hear in my head today. Whenever I reacted as if the world was coming to an end because, for example, I had been passed over for a singing part in the Christmas pageant in second grade, she'd roll her eyes and say, "Samantha, don't make a mountain out of a molehill."

In therapy, I was trying to do the opposite. My goal was to reduce mountains into molehills.

Velcro people, like Gina, did not take to this message. Often they would respond to a challenging or ambiguous situation by magnifying its negative potential. Psychiatrists call this catastrophizing. For example, imagine there's a female catastrophizer applying for a job when the interviewer stops abruptly after twenty minutes, thanks her for coming, and walks out the door. A catastrophizer might respond by believing (a) the interviewer absolutely hated her, (b) she bombed the question about what her weaknesses were, and that (c) she must have said something terribly wrong to offend the interviewer. A spiral of negative thinking ensues until finally she believes that she will never get this job—or *any* job, for that matter.

The interviewee imagines the worst-case scenario so vividly, she becomes convinced it actually happened and doesn't send a follow-up email to the company expressing interest in working there. Crestfallen, she also doesn't attend the company roundtable for potential employees that evening. You get the picture.

During my training I learned that the best way to challenge catastrophizing was to help the patient recognize that their beliefs were not necessarily true and to generate alternative interpretations. Although a bad interview was certainly a possibility, there are other plausible explanations for the meeting having ended abruptly. Perhaps interviews are slotted to last only twenty minutes. Perhaps the interviewer's cat had an emergency hairball. Perhaps the interviewer was so impressed that a quick conversation was enough.

It was the *interpretation* of the incident—not the incident itself—that triggered the cascade of catastrophic thoughts and the snowballing of negative emotions. The point was to help a patient see that her perception might be distorted and that the mountain she faced was, in fact, a molehill, or at least more of a Mount Pleasant than a Mount Everest.

Focusing on an individual and his or her problems has value, but if that's all psychiatrists do, we're leaving a lot on the table. How fortified the patient feels matters, too. When a Harvard School of Public Health report asked people what helped them feel strong in the face of stress, the respondents pointed to activities that gave them a boost. Prescription medication and professional help made it onto the list but hovered near the bottom.

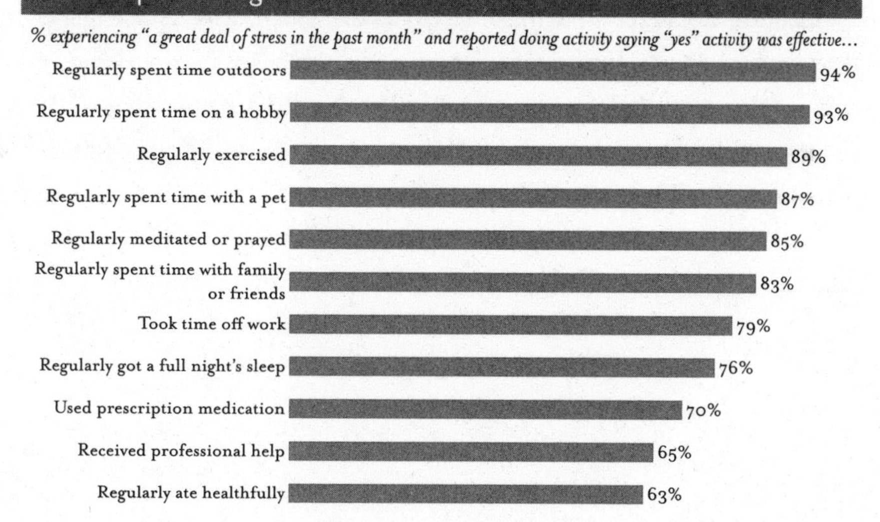

Harvard Opinion Research Program Survey Series. The Burden of Stress in America *(conducted by Harvard School of Public Health from March 5 to April 8, 2014, in partnership with NPR and the Robert Wood Johnson Foundation).*

Regularly spending time outdoors topped the list, followed by spending time on a hobby and exercising regularly. Other fortifications cited were getting a good night's sleep, eating well, and spending time with family and friends. Whether introverted or extraverted, people usually get a boost from being with others.

Other proven mood-enhancing boosters from the study included

- doing something for others;
- contributing to something beyond oneself;
- learning something new;
- doing something creative;
- moving around or exercising; and
- doing what one does best. (The more hours in the day that a person uses their strengths, the less likely the person is to report worry, stress, or sadness.)

Overall, activities that make "a good day" are active and engaged; they are less about individual reflection and more about participating and connecting. A great deal of everyday well-being lies beyond what is happening inside a person's head. Everyday opportunities and activities that foster growth and build positive resources are not "icing on the cake" but the active ingredients of everyday resilience. The *r* might be little, but the effect is not.

Chapter 4

PEOPLE CHANGE

> Human beings are works in progress that mistakenly think they're finished. The person you are right now is as transient, as fleeting, and as temporary as all the people you've ever been. The one constant in our life is change.
>
> DANIEL GILBERT

Most people want to make some sort of change in their lives. That's why they step through the door of my office, which is not always an easy thing to do. But as therapy progresses, patients often hit a wall when faced with the challenging process for achieving actual change. They're often interested more in other people changing than in bringing about any changes within themselves.

"I am who I am," they stubbornly insist.

I've heard that line countless times, along with the idiom, "You can't teach an old dog new tricks." These declarations are evidence of a fixed mindset and the flawed belief that our abilities, skills, and characters are set in stone. Society labels us, and we in turn label ourselves. A child is told he is a good or a bad listener. A high school student thinks of herself as good or bad at math. By the time we're adults, these labels become deeply ingrained into our sense of self and part of our identity or self-concept.

Believing something, however, doesn't make it true. My dog Schnitzel recently discovered her talent for jumping through hoops, so you *can* teach an old dog new tricks. I once tried to convince a self-described workaholic to use her vacation days by the end of the year. "I am not that kind of person," she replied. Being raised by a single mom had given her an uncompromising work ethic. If she wanted to succeed, her mother had always told her, she would have to outwork everyone else. This gave her a fantastic work ethic—along with an overly rigid sense of self.

Like this workaholic patient, many become strongly attached to ideas of what they are *not*:

> *I'm just not a relationship guy.*
> *I'm not a people person.*
> *I'm not an athlete.*
> *I'm not a morning person.*
> *I'm not the type who apologizes.*

These are all lines I have heard from patients, often followed by the phrase, "It's just who I am," accompanied by a shrug of the shoulders to punctuate this conviction. Such statements are intended to project self-knowledge, but they are, in fact, just tightly held beliefs; the product of a lifetime of habits, choices, and the experiences that reinforce them.

Patients aren't the only ones resistant to the concept of change. Up until the early 1960s most scientists believed that the adult brain was completely formed and inflexible. Any possibility of neuroplasticity—the brain's ability to form new connections—and neurogenesis—the brain's ability to regrow neurons—was dismissed as fantasy. Today, we know that the brain continues to reorganize and respond to changing needs and situations throughout our lives.

What happens to the brains of London taxi drivers provides a fascinating example. Unlike those in other cities, London cabbies are required to memorize thousands of street names and routes to pass a notoriously difficult licensing exam known as "the Knowledge." In 2011, researchers at University College London decided to investigate the impact of this training on the brain. They scanned the brains of Knowledge candidates before they embarked on the lengthy training process and then again three to four years after qualifying for a license. Imaging studies revealed that the drivers' hippocampi—the part of the brain associated with memory—had become significantly larger after mastering the map of London. Similar growth was not observed in the brains of those who failed to qualify or in people who were not studying for the Knowledge. Learning something new creates new connections in the brain. These pathways are formed to help us adapt and respond to challenges and whenever we are exposed to unfamiliar situations and environments.

The fact that the brain has a tremendous capacity to rewire itself underscores the potential we all have to change, to reimagine, and to reinvent ourselves throughout our lives. A study that spanned the course of sixty-three years revealed how dramatically these changes can be. Researchers found a group of Scottish elderly who had been evaluated by their teachers in 1947. Their teachers had been asked to rate the students, who were fourteen years old at the time, on six personality characteristics: self-confidence, perseverance, mood stability, conscientiousness, originality, and desire to excel. More than six decades later the researchers asked the seventy-seven-year-olds plus a friend or relative to rate them on those same characteristics. The overlap was minimal: who they were at age fourteen had little to do with who they were at age seventy-seven.

Letting go of the notion that people "are as they are" can help reduce depression and anxiety. High school students who were given lessons on the topic of neuroplasticity and who learned that personality traits are not fixed were better equipped to handle stress, were more confident, and achieved better grades than those who didn't. Individuals open to the idea of growth are the people most likely to grow. They believe that interests and passions are not innate but can be developed.

Life experiences clearly shape our personalities. When psychologists discuss personality traits, they are often referring to the "Big Five": openness to experience, conscientiousness, extraversion, agreeableness, and neuroticism. The particular combinations of these five traits are believed to be the core characteristics that form an individual's personality.

Strong relationships increase conscientiousness, agreeableness, and extraversion and decrease neuroticism. Greater job satisfaction decreases neuroticism and increases extraversion. We can also actively change ourselves by consistently engaging in behaviors that reflect the traits we would like to have. For example, in one study, people who said they would like to be more extraverted took intentional steps to behave in an extraverted fashion, such as to smile at someone new, ask a cashier how their day is going, and call a friend they hadn't spoken to in a while. During a three-month period they became more extraverted. Those who said they would like to be more agreeable and took concrete steps to behave in an agreeable fashion, such as to hold the door open for someone, to say "please" and "thank you" when asking for something, and to give a friend or family member a sincere compliment, likewise became more agreeable over time. Leading a physically active life can also impact your personality. Active people tend to become more conscientious, open, extraverted, and agreeable. The opposite tends to be true for couch potatoes.

The older we get, the harder it becomes to imagine ourselves being any other way. As journalist Lindsay Crouse points out, "I had always thought that, at some point in life, most people become 'who we are.' Our lives are built around whatever that is, and no matter what we might in fact be capable of, this idea keeps us fixed in one place. At 35, I thought I was 'who I was.' I didn't think it was still possible to improve significantly in anything, let alone something involving my body."

Crouse decided to prove herself wrong by training to do "the impossible"—to run a marathon faster than she ever had before and to qualify for the Olympic marathon team trials. Her time of two hours and fifty-three minutes fell just short of qualifying but far exceeded what she thought her body could achieve. It is not just through experience but also through application that we gain control over the kinds of people we are and can become.

Just as it's important to recognize your own potential to change and grow, it's also important to allow for change in others. Too often we judge people quickly and interpret a minor offense as a permanent character flaw. Young people are especially vulnerable to jumping to conclusions. From getting bumped in the hallway or left out of a game during recess, students determine that the "perpetrators" are bad people, acting purposefully to do harm to others. Students who are taught that individuals can change are less likely to think this way and, more important, they're less likely to want to retaliate. Placing others in categories may make the world seem more predictable, but it limits us from seeing them fully.

A major challenge for all of us is to refrain from making assumptions about others based on only partial knowledge. Our brains have a tendency to place people and things into categories. A patient once explained to me during an initial evaluation, "The way I see it, people are either with me or against me." When I responded that her treatment would involve learning to question knee-jerk responses and to resist the impulse to dismiss or judge based on kernels of information, she decided I was not the right psychiatrist for her.

"Isn't it the psychiatrist's job to be on the same team as the patient?" she asked.

The psychiatrist's job, I explained, is to encourage a patient to appreciate complexity and nuance, and to discourage the viewing of life through an "us and them" prism.

Creating categories is useful when it helps the brain sort through the vast stimuli of everyday life. Every time we encounter a waist-high object with a flat surface and legs, we can assume it's a table—be it a dining table, a side table, or a card table. This saves time and energy. But the tendency to categorize other people can lead to reductionist assessments that inhibit understanding and reduce empathy. Stanford's Jennifer Eberhardt explored the roots and ramifications of bias and found that people subconsciously associate crime-related objects, such as a gun, with Black faces and that teachers were quicker to see patterns of bad behavior in Black children. Stereotypic associations like these give rise to both baseless assumptions and racial discrimination. Whenever you find yourself assuming malintent or are unwilling to give someone the benefit of the doubt, pause for a moment and check your assumptions.

Let People Surprise You

It is often assumed that cocaine or heroin carry the most life-threatening withdrawal symptoms, but actually alcohol withdrawal carries the greatest risk of death. When I met Pete in the intensive care unit (ICU), he was at risk of developing delirium tremens, a severe form of alcohol withdrawal. Delirium tremens typically develops in chronic alcohol abusers two or three days after they stop drinking, and can lead to seizures, respiratory failure, and fatal arrhythmias. Without treatment, more than a third of those with the DTs will die.

I was on the CL (consult-liaison) service at the time, the team that treats patients on the medical wards who also have a psychiatric diagnosis. Pete was a chronic substance abuser who had been in and out of the hospital for years. He had not been drug- or alcohol-free for more than a few days in over a year, so we followed his vital signs closely and assessed his neurologic status frequently. We stabilized him with IV fluids and Librium. After a few days he was healthy enough to be discharged from the ICU, although I didn't hold much hope for him maintaining his sobriety, because he'd been circulating through the revolving door of rehab and relapse.

The next time we crossed paths, Pete had been sober for two years, was employed full time, and engaged to be married. It was hard to reconcile the image of the ailing patient I knew from the ICU and the robust man now standing before me. Sensing my bewilderment, he looked at me with a smile and said, "People change."

It is hard to pinpoint what led to Pete's transformation. When I first met him, he exhibited the belief known as terminal uniqueness, a grandiose conviction that he was fundamentally different from others who were dealing with substance abuse. He genuinely believed that he had suffered more, that his circumstances and experiences were special, that, unlike other people who enter Alcoholics Anonymous, he didn't have to abstain from alcohol, and didn't require the same kind of treatment others did. The "terminal" part of "terminal uniqueness" refers to the fact that this sort of thinking can ultimately be fatal. My professional opinion is that overcoming his terminal uniqueness must have been a big part of Pete's recovery. Witnessing growth and recovery are two of the most satisfying parts of my profession.

Beatrice

In my second year of training I met Beatrice. She was twenty-three years old and wore her hair in two shoulder-length braids with big bows, which made her look like Cindy Brady from *The Brady Bunch.* She walked around the inpatient unit wearing oversize pajamas and fluffy slippers. Sometimes she would suck her thumb. When she sat in a chair, she would curl her legs up tightly beneath her, which made her appear to shrink before my eyes—a position that is known as "the borderline scrunch."

Borderline patients are characterized by mood swings, self-destructive behavior, and unstable relationships. They often come across as immature for their age and tend to see people as either good or bad. One minute they may be filled with appreciation—"You are the best psychiatrist I've ever had"—while the next they may be threatening to report you to the medical board. This rollercoaster pattern of neediness and rejection plays out with family members, friends, and romantic partners. In general, psychiatrists do their best to avoid hospitalizing patients with borderline personality disorder because they tend to deteriorate in that setting, becoming more regressed and less capable of managing their emotions. Still, sometimes hospitalization can't be avoided.

Beatrice was on her fourth hospital stay when we met. During a row with her mother, she had grabbed a bottle of vodka and locked herself in her bedroom, shouting that she was going to use the liquor to wash down a bottle of Tylenol. Panicked, her mother called 911. The paramedics arrived, knocked down the door, and found Beatrice lying on her bed with an empty bottle of pills next to her. She was responsive and told them she had taken a handful of Tylenol. (Her bloodwork later showed no signs of acetaminophen toxicity.) When she arrived at the ER, the staff administered a slurry drink of charcoal to absorb whatever toxic substance she might have ingested, and then her outpatient psychiatrist turned her over to our care. She had blown through a number of doctors ("They sucked, so I fired them") and was ready to dispense with the new one as well. On the unit she caused problems. She flirted with a young male patient and was caught trying to sneak into his bedroom one evening, and she accused the staff of stealing patients' cigarettes.

I found Beatrice a new therapist and discharged her to an intensive outpatient program. The most effective treatment for borderlines is dialectical behavior therapy (DBT), which focuses on building coping skills to help patients manage their symptoms. Still, it's hard to be optimistic about the prognosis for a low-functioning case of borderline personality. I had a pretty good sense of how things would turn out, and my go-to thought in such cases was, "I know the end of this movie." I assumed Beatrice would be hospitalized again and again. There would be more suicidal gestures, and a comorbid substance abuse was likely—probably alcohol, maybe benzos. She would continue living at home with her mother, who would also bear the burden of her daughter's mental illness.

Friendships would drop off, and Beatrice would become increasingly isolated and dysfunctional.

A few years later I read a newspaper article about how Beatrice had started a successful company in the food industry. In the photo with the piece she was accompanied by a child in a stroller and her partner. A ponytail had replaced the Cindy Brady braids. The article mentioned her early struggles with mental illness and how DBT had changed her life. This particular movie had a twist at the end, and my being wrong never felt so good.

When we are so convinced we are right, we fail to see other possibilities or to appreciate endings. As a hedge fund manager explained to me, "There is a thin line between arrogance and confidence. . . . It's called humility."

Change by the Bundle

Whenever people want to make major changes in their lives, conventional wisdom holds that they should focus on one thing at a time. The thinking goes that if they try to make too many changes at once, they will overload and fail. Patients tell me that only after they stop smoking will they start getting in shape, or once their anxiety is under control, then they will start eating healthfully. This "step-by-step" mindset seriously underestimates our ability to make multiple simultaneous changes in our mental and physical health. In fact, a comprehensive approach actually enhances success in *all* areas. In one study, a group of healthy young adults who worked concurrently at getting more sleep, drinking less alcohol, exercising more, improving relationships, being more mindful, and eating healthier showed dramatic improvements in a variety of domains, including strength, endurance, flexibility, memory, standardized test performance, focus, mood, and self-esteem.

Lead researcher Michael Mrazek explained, "Recent research suggests it's often more effective to make two or more changes simultaneously, especially when those changes reinforce one another. It's easier to drink less coffee if at the same time you get more sleep. Our intervention extended this logic by helping people make progress in many ways, which can create an upward spiral where one success supports the next." The results of the study were described as "clear and striking" *and* they lasted. Even six weeks later participants continued to show improvement in all areas. This research makes me feel optimistic about what is possible for all of us.

Philosopher and psychologist William James says it best: "To change one's life: 1. Start immediately. 2. Do it flamboyantly. 3. No exceptions."

PART TWO

Choose Vitality

Chapter 5

HARD, BUT IN A GOOD WAY

In one of my favorite experiments in modern psychology, rats were divided into two groups. Each group was given a Froot Loop—an obvious rat delicacy—every day for five weeks. For one group, the Froot Loop was buried under a mound of bedding in their cage. These rats needed to search for and dig up the treat. Kelly Lambert, professor of behavior neuroscience at the University of Richmond, called them "worker" rats. In another cage, rats were simply offered the daily Froot Loops. These rats were dubbed "trust fund rats," because they basically received a Froot Loop on a silver tray each day, minus a butler in white gloves.

After a few weeks, both sets of rats were presented with a new challenge: the treasured Froot Loop was placed into a clear plastic ball. The rats could see and smell the treat, but could not get to it. (For me, it would be the equivalent of being given a chocolate fudge brownie in a locked glass jar.) The reactions from the worker rats and the trust fund rats were notably different.

The worker rats tossed the ball from side to side, hurling it across the cage. They tried to stick their paws into the openings of the ball to reach the treat. Undeterred and undaunted, these little gladiators refused to give up even in the face of frustration. The trust fund rats? Not so much. Although they clearly wanted the Froot Loop, they didn't put in nearly as much effort as the worker rats. In fact, they made 30 percent fewer attempts and spent 60 percent less time trying to get the treat.

Lambert theorizes that the worker rats were bolder and more persistent when presented with this new challenge because of the effort they put into the original one. Having to dig up the Froot Loop taught them that with effort, they could overcome difficulty. In psychology jargon, working for the Froot Loop cultivated rat "self-efficacy"—a belief in their ability to succeed. In a separate experiment, a group of rats in Lambert's lab tackled a different challenge and learned to drive tiny plastic cars—ROVs (rat-operated vehicles)—in pursuit of the Froot Loop. Learning to drive was somewhat stressful, as evidenced by the amount of stress hormone corticosterone detected in the feces of the motorist rats. But the new skill also led to the increase of DHEA (dehydroepiandrosterone), a hormone that counteracts stress and is a recognized marker of emotional resilience. Acquiring a new skill and achieving mastery cultivated rat resilience.

"Anything that lets us see a clear connection between effort and consequence—and that helps us feel in control of a challenging situation—is a kind of mental vitamin that helps build resilience," Lambert observed.

Taking a page out of the rat playbook, the more effort we apply to accomplishing a difficult task, the more we build a reservoir of experiences that embolden us to tackle future challenges and respond flexibly to novel situations. When it comes to humans, Lambert is concerned about how disconnected our day-to-day lives have become from effort-based experiences.

An undemanding existence is certainly convenient, but is it eroding well-being? A life of ease and comfort masquerades as the ultimate prize, but is it depriving us of everyday opportunities to build resilience by taking the easy way out? Are we choosing immediate gratification and indulgent self-serving pleasure over a more engaging and ultimately more rewarding challenge?

Desirable Difficulty

In 1915 a German warship torpedoed the British ocean liner *Lusitania* off the coast of Ireland, sinking it and drowning almost twelve hundred passengers, including 128 American citizens. A hundred years later my son had to deliver a report on this attack and its significance to his sixth-grade class. It was a challenging assignment, and he spent stressful weeks rehearsing his presentation, knowing he would be judged by his peers. His preparation paid off, and when I asked him afterward about the experience, he responded with a smile, "It was really hard, *but in a good way*."

In theory the idea of maximizing pleasure and minimizing discomfort whenever possible makes sense, but the reality is that a task can be hard, and challenging, and even stressful . . . *but in a good way*.

Desirable difficulty is not a paradox. Robert and Elizabeth Bjork's research demonstrates that students are more engaged and learn more effectively overall when they have to figure things out for themselves. Being spoon-fed material that is too easy or straightforward is boring and easily forgotten. I had a friend in school who spoke fluent Spanish but got a grade of B in Introductory Spanish because she was simply bored out of her mind in the class. The right amount of challenge draws students into the learning process and invites deeper processing, which makes the knowledge stick. For example, when studying for a test, students typically opt for the easy strategy of reviewing their notes or rereading the textbook. But research results indicate they'd be much better off taking the more stressful practice tests. Even if the students get some of the answers wrong, the process of looking up the correct one better embeds the information into memory.

Stress has a bad reputation, but studies show that some stress is in fact good for us. Hans Selye, "the father of stress research," demonstrated that good stress—or eustress—is a powerful motivator. Eustress pushes us to do our best and facilitates optimal performance. Too little stress—hypostress—can lead to boredom and feelings of ineffectiveness and even depression. Of course, a given challenge and the stress that accompanies it are beneficial only if we have the resources available—including the ability, stamina, energy, and time—to meet it.

With so much focus on happiness these days, it's tempting to look for shortcuts to bypass challenges of any kind. Unfortunately, shortcuts rarely work, and there is a lot to be gained by taking the long cut. Researchers at Harvard and Duke asked one group of students to assemble an IKEA storage box, while others were given fully assembled units. Both groups were then asked which boxes they liked more and what they would pay for them. You might think that having to build the box would drive the price down and make them less appealing, but it seems the builders preferred their own boxes to the preassembled ones and were willing to pay 63 percent more for them. The researchers called this phenomenon the "IKEA effect," in honor of the Swedish manufacturer whose products typically require some assembly; "When Labor Leads to Love" is the title of their paper about the experiment. Building the boxes made people feel competent, capable, and proud. Anyone who has ever put together a piece of furniture can relate. Even though the completed chair might be a little wonky and one of its legs might be askew (and ignore that chip in the paint because the screwdriver slipped), the fact that you made it with your own hands adds value.

Hands-on activities engage us. To prove this point is a story often cited in marketing psychology classes, even though its origins are murky. Still, the story resonates and involves cake, so I think about it quite a lot. In the 1950s General Mills introduced a new line of prepackaged cake mix that was, well, as easy as pie. The box contained all the necessary ingredients—flour, sugar, and milk and eggs in powdered form. All a busy housewife needed to do was add water, stir, pour the batter into the pan(s), stick it into the oven, and voilà, a delicious home-baked cake was ready to serve. Executives at General Mills were confident the product would be a huge success and transform the way women baked.

Except . . . the state-of-the-art cake mix was a total flop. General Mills brought in a team of psychologists to investigate why this product didn't have the anticipated appeal. The problem, according to the team, had nothing to do with the cake mix itself. It was the *process* that just didn't *feel* right. Bakers appreciated the ease and speed of using the cake mix, but not having to invest any effort into the process left them feeling sidelined and empty. Contrary to conventional marketing wisdom, the product failed because it was *too* convenient.

The solution, so the story goes, was for the food scientists to rework the recipe. They removed the powdered egg and required the cake maker to supply a real one. The physical act of cracking an egg and stirring it into the mix meant the baker became a greater part of the process.

Legend has it that when General Mills relaunched the product with an "Add an Egg" slogan, the cake mix flew off the shelves—a perfect example of how the joy of experience beats the cult of convenience. Tim Wu, an author and law professor at Columbia Law School in New York, summed up this principle beautifully in a 2018 op-ed, writing, "Convenience is all destination and no journey. But climbing a mountain is different from taking the tram to the top, even if you end up at the same place. We are becoming people who care mainly about outcomes. We are at risk of making most of our life experiences a series of trolley rides."

My personal favorite example of the IKEA effect combined with an Add an Egg contribution comes from my own kitchen. My daughter was just learning to bake and had spent the morning toiling over a mixing bowl. Hours later she sat at the table, eating a piece of yellow butter cake. I asked her how it had come out. "This is the best cake I have ever eaten," she declared, reveling in the fruits of her labor, savoring every bite. The cake certainly did taste good, but to her, it was the most delicious baked good in the history of confections. Why? Because she had made it.

Look for opportunities to "add an egg" in your daily life. We tend to overvalue fleeting pleasure and forget about the enduring satisfaction that accompanies engaging in activities that are difficult but also satisfying.

Being at work and spending time with children are rated among people's least daily pleasurable activities, but they are also rated as among the most rewarding. Ultimately, we are not so different from those worker rats.

Stressful . . . But in a Good Way

Plenty of people with mental illness have managed to find what Dr. Elyn Saks calls "wellness within the illness." Dr. Saks isn't just an objective observer. While a student at Yale, she was diagnosed with schizophrenia. Her parents were advised to recalibrate their expectations for her, to take her out of school, and to remove any possible stress in their daughter's life. They were told that it was unlikely she would ever live independently, hold down a job, or get married. Saks and her family accepted the diagnosis of schizophrenia, but refused to accept the prognosis of a life defined by it.

"I was told to get a job as a cashier at a store," Saks said in an interview. "I thought to myself, I'm a student. I'm good at it. I like it. . . . What was more stressful to me is the idea of a constant line of people asking for change."

With therapy, medication, and tremendous support from family, Saks persevered. She graduated from college and then law school. Today she is a champion of mental health law and a highly respected professor and associate dean at University of Southern California's Gould Law School. She is married. Her frank and poignant memoir, *The Center Cannot Hold: My Journey Through Madness*, detailed her struggle with both mental illness and misguided treatments.

"Using my mind—is my best defense," Saks has said. "It keeps me focused; it keeps the demons at bay. . . . Far too often, the conventional psychiatric approach to mental illness characterizes people as clusters of symptoms. Accordingly, many psychiatrists hold the view that treating symptoms with medication is treating mental illness. But this fails to take into account individuals' strengths and capabilities, leading mental health professionals to underestimate what their patients can hope to achieve in the world."

Eliminating stress was not the key to defying expectations for Saks. Dropping out of school would have eliminated the immediate pressure, but it would also have diminished her potential. It was engagement with the world that led to her success. Psychiatrists sometimes forget that what is meaningful is often stressful. It is important not to deprive patients of the experiences they value and the full potential for which they strive. Our duty is to better enable them to imagine what's possible and to help them reach their goals and live their lives with vitality.

In 2009 Saks was awarded a MacArthur fellowship for her work as both a legal scholar and advocate for mental health policy. The Foundation commended her for "expanding the options for those suffering from severe mental illness through scholarship, practice, and policy informed by a life story that adds uncommon depth and insight."

The Joy Is in the Doing

Psychiatrist Richard Friedman, who runs the mental health services for Weill Cornell Medicine of Cornell University in New York, has voiced concern that the emphasis on wellness today is creating unrealistic expectations that everyone should be smiling and stress free at all times. "Though I can't prove it," Friedman wrote, "I suspect that my generation suffered less burnout than current students for the simple reason that we expected to have a rough ride."

Each year Friedman greets incoming first-year medical students by telling them, "These next four years will be exciting and challenging and stressful. . . . It's entirely natural to feel anxious, overwhelmed at times, and exhausted. In fact, it's evidence you are alive and engaged in your work." We feel stress because we care and because we're striving. Poet David Whyte describes this succinctly: "In romance, in parenting, and in our professional lives—when we're fully committed and deeply engaged, we get hurt, feel frustration, upset. And that's not a bad thing. It means you are sincere and committed."

During medical school, I used to imagine how happy I'd be after taking the first in a series of medical licensing exams. I had studied for weeks, taking countless practice tests and going over the difference between interstitial cystitis and schistosomiasis. In my post-exam fantasy I literally pictured myself dancing in the streets, kicking up my heels like they do in old movies. I was certain the euphoria would last for weeks.

The reality was different. When the exam was over, I felt more relief than outright joy. There was no dancing in the streets. A few weeks later I received a letter informing me that I had passed. I was elated, but again, that feeling didn't last long. Almost immediately I started to wonder, *What now?*

I didn't miss the grind of studying for exams, but I missed the *engagement*. The process of mastering the material was purposeful. It was meaningful to me to learn the material thoroughly, and I hoped it would be meaningful to my future patients. In retrospect I'm not convinced that any of them actually benefited from my knowledge of how many ATP molecules are produced during anaerobic respiration, but the experience of learning that fact was itself worthwhile to me.

Psychologist Richard Davidson describes two kinds of goal-related positive emotions. One is called "post-goal attainment"—that short-lived feeling of contentment upon reaching a goal, like I experienced when I first completed and then passed the medical exam. The other is called "pre-goal attainment"—the positive feeling that comes from moving toward a goal. *That* was the feeling I missed experiencing after I took the test. In his book *The Happiness Hypothesis*, Jonathan Haidt explains how the experience *is* the reward:

When it comes to goal pursuit, it really is the journey that counts, not the destination. Set for yourself any goal you want. Most of the pleasure will be had along the way, with every step that takes you closer. The final moment of success is often no more thrilling than the relief of taking off a heavy backpack at the end of a long hike. If you went on the hike only to feel that pleasure, you are a fool. People sometimes do just this. They work hard at a task and expect some special euphoria at the end. But when they achieve success and find only moderate and short-lived pleasure, they ask (as the singer Peggy Lee once did): Is that all there is?

Like any good poet, Shakespeare summed it up in far fewer words: "Things won are done; joy's soul lies in the doing."

Chapter 6

BE UN-YOU

Stop Worrying about Finding Yourself

Not long ago I was invited to give a talk at the American Psychiatric Association's annual meeting. When I realized I couldn't back out, I went into full panic mode.

"Just go out there and be yourself," suggested a well-meaning colleague.

I smiled and thought, *Well*, that's *terrible advice.*

Public speaking has always filled me with dread, and I was pretty certain that simply being myself would lead to my either collapsing at the podium or escaping through the back door. What I needed to be was "un-me."

An old friend told me about a patient who shared my performance anxiety and recommended a counterintuitive strategy. Before facing an audience, the patient explained, "My heart starts racing, I feel like I can't breathe, beads of sweat collect on my forehead, my hands are shaking, my palms are sweating, and I feel sick to my stomach." One night he was watching a late-night talk show on which Bruce Springsteen was a guest. The host asked the Boss how it felt to go on stage and perform in front of twenty thousand people. Springsteen responded: "It's the most incredible feeling. I feel my body kicking into high gear. My heart starts racing, I start breathing a little harder, my palms are sweating, my hands are shaking, I feel sweat on my brow and I have butterflies in my stomach. It's a sign to me that my body is ready to rock."

Both men's physiological symptoms were strikingly similar—elevated heart rate, sweaty palms and forehead, rapid breathing—and yet their interpretations of them were radically different. The patient realized that his problem wasn't performance anxiety, but rather his inability to get out of his own head. From then on whenever he had to speak in public, he thought of himself as the Boss. It helped him be un-him.

I needed to do the same and pretend to be someone who spoke well, and who was accustomed to being in the limelight. The answer for me was obvious: Barbara Walters. The acclaimed anchorwoman had recently given a speech that I watched in awe. She was confident, self-assured, funny, and unflappable—everything that I needed to be.

On each page of my speech, I wrote the initials "BW" to remind me to stay in character. I adopted her posture and imagined how she would look out from the podium and smile at the audience. I spoke slowly and with conviction.

For the first time I delivered a good speech. Instead of escaping out the back door, I escaped the wave of insecurity that would have enveloped me had I been myself. Today, whenever I give a speech, I still scribble her initials on my notecards as a reminder.

Be Someone Else

There is evidence that looking beyond oneself and channeling someone whom you admire provides better guidance than stewing in your own emotions. A study of children highlights the benefits of not being yourself. A group of six-year-olds was asked to work on a repetitive task on a laptop but could take a break whenever they wanted to play games on an iPad. The iPad was placed right next to them. One group of children was told to think about their own thoughts and feelings. A second group was told to think about themselves in the third person. A third group was told to think about someone else who was really good at working hard and to pretend to be them. Batman, Rapunzel, Dora the Explorer, and Bob the Builder were possible choices. The iPad games proved to be a tempting distraction for all the kids, but the kids who pretended to be someone whom they admired persevered the hardest and staved off temptation the longest.

I am not suggesting you go out and buy a Batman costume—okay, maybe I *am* suggesting that—but this research has relevance for how we face challenges and hassles.

Pretending to be someone else can promote flexibility in the midst of anxiety. One patient found inspiration in the Cold War spy thriller *Bridge of Spies*. In the film, lawyer James Donovan, played by Tom Hanks, defends Russian spy Rudolf Abel, played by Mark Rylance. While preparing for the trial, Donovan is baffled by Abel's nonchalance and cool-as-a-cucumber demeanor, given the serious charges against him and likelihood of a death sentence.

"You don't seem alarmed," Donovan observes.

Abel responds nonchalantly with a reasonable question: "Would it help?"

This scene resonated with my patient. He knew that panic doesn't help and can get in the way of thinking clearly and rationally. So when he landed in a situation that he could not control, he'd channel his inner Soviet spy. Looking to "emotional exemplars" helps access resources like self-control, persistence, confidence, and creativity when your own are depleted. This behavior is different than self-distancing—it's self-broadening. Approaching a challenge from another person's point of view better enables you to view it with fresh eyes. It is emulation, not imitation. Or as author Adrienne Brodeur wrote in her memoir, *Wild Game*: "You have no idea how much you can learn about yourself by plunging into someone else's life."

I have one patient who was required to testify in a court case involving a former boss who had been accused of malpractice. She was absolutely terrified about what the lawyers would ask her. What got her through the experience was channeling the unflappable Hillary Clinton during the Benghazi hearing. A different patient found a way to deal with a tone-deaf colleague by asking herself, *What would Oprah do right now? Oprah would speak up!* Of course, the role models need not be celebrities. I know one young woman who thinks of her grandmother whenever her patience is running low. It helps her access the empathy and understanding that might be hard for her to find in a particular moment.

Tapping into the capabilities of those who exemplify qualities or abilities we wish we possessed may, in fact, help us find them for ourselves. A study found that people demonstrated greater flexibility and were more successful at creative problem-solving when they imagined themselves to be eccentric poets. When people typically think about creativity, they assume it is a fixed trait, a talent people are either born with or not. But as this study highlights, to unlock creativity we may only need to get out of our own head and imagine ourselves in that of a creative individual.

Change Your Story

Today there is a great deal of emphasis on authenticity—meaning that your outward behavior must match your internal feelings. The implication is that unless you're embracing your natural inclinations and expressing your "true self" at all times, you are a fraud or what my stepson, Charlie, would call, a NARP—not a real person. But this is a limited way of seeing yourself and denies the possibility of expanding beyond yourself and becoming something more. Believing there is a singular fixed or true self can interfere with growth and has been linked with depression.

When I was training to become a psychiatrist, an old master in the field asked our class what I thought was an obvious question: "What do you think the point of therapy is?" Eager beaver that I am, my arm shot up. "The point of going to therapy," I said confidently, "is to give yourself a brighter future."

"Wrong, Dr. Boardman," the professor snapped back. "Anyone else?"

Another brave resident jumped in and offered, "The point of therapy is to change your present."

"Wrong again!" the professor bellowed. "The point of therapy isn't to change your present or your future. The point of therapy is to change your *past*."

People lock into their own narratives. Their older brother was the favorite. Their first boyfriend was the love of their life. Their parents were not supportive. These stories are told and retold so many times that they become canon. All too often a person's sense of worth and identity is based on these beliefs, which do not take into account nuance and detail. Letting go of the established stories we tell about ourselves allows us to decide who we want to be, instead of letting the past dictate who we are today and who we will become in the future.

Breaking free from the idea of a fixed self can set us free in ways big and small. A few summers ago my children were invited to a trapeze party. The instructor asked me to join. "No thanks," I instinctively responded. Trapeze is not me." But my children insisted. I climbed up to the narrow platform, put my toes over the edge, clung on to the bar quite literally for dear life, and stepped off.

It was a terrifying and completely humiliating experience, thanks to an undignified dismount that left me face-planting into the net below. And it was worth every thrilling second, for it was completely out of character. Doing the "un-me" has saved me on many occasions from a self-defeating reflexive response that might have felt more "me" but prevented me from enjoying an experience or an interaction or an opportunity to learn something.

Who we think we are can get in the way of growth and vitality. Rather than to be their "true selves," which tends to narrow perspective and promote rigidity, I encourage patients to expand how they think about themselves by behaving in ways that may seem out of character.

"I'm a 'yes' person," a patient once told me. Being nice was very important to her, and she was known by her friends to be agreeable and accommodating. Because of this reputation, though, she sometimes felt used and taken advantage of but was afraid to behave any differently. "Being nice is who I am," she insisted, while deep down, she believed that her "niceness" was the only reason people liked her. I asked her to consider the difference between what it means to be nice and what it means to be kind. We talked about how "nice" is about being a pushover, never saying what you think and doing what other people want you to do, while "kind" is about staying true to your values and exhibiting the grace and strength to express yourself. A few days later, a coworker asked her to stay late to finish a PowerPoint presentation. In the past, she would have said yes, even if she had plans. This time she decided to be un-her and declined.

Acting out of character enabled her to rise out of the confines of her limited self and helped her to find her voice, to feel more confident, to be a better version of herself. I don't think of this technique as denying or disrespecting who someone is or asking a person to be inauthentic. On the contrary, it helps that person get closer to the version of themselves they would like to be. Is she being inauthentic? Technically, yes. Still, being a well-meaning phony can sometimes be the key to self-transformation.

What does it really mean to be authentic, anyway? Scientists from UC Berkeley asked people what was more important for feeling authentic in a romantic relationship—being your actual self, or being your ideal self. The majority assumed that being your true self was the key to an authentic relationship. But research tells a different story. Authenticity in a relationship is the result of feeling you can be your best self, not your actual self.

Doing things that are "un-you" can free you from behaving in a way that may be comfortable but stifling. Disagreeable people feel better when they are more considerate. People who are careless feel better when they are conscientious. Shy people feel better when they act more outgoing. Psychologist Sonja Lyubomirsky demonstrated that when introverts intentionally engage in extraverted behaviors, such as being assertive, talkative, and spontaneous, they can increase feelings of connectedness and gain an overall boost in well-being.

I have always asked my children to look for emotional exemplars in the books they read and the movies they watch, though I was puzzled when my daughter told me that one of hers was a spider: "When I need a little help thinking something through or have an issue with a friend, I sometimes ask myself, What would Charlotte do? Unless I'm trying to decide what to eat, it's really helpful."

It took me a moment to realize she was referring to Charlotte, the wise and selfless spider in E. B. White's *Charlotte's Web.* Charlotte's farewell to Wilbur captures the beauty of friendship and connection: "After all, what's a life anyway? We're born, we live a little, we die. A spider's life can't help being something of a mess with all this trapping and eating flies. By helping you, perhaps I was trying to lift up my life a trifle. Heaven knows, anyone's life can stand a little of that." Whether it's a spider or a spy, inspiration to be the best version of ourselves most often comes from outside of ourselves. Engaging with complex characters with rich inner lives helps us envision alternative responses to dilemmas and expand our repertoire of possible reactions. We wonder what we would do in their situations. It can also be helpful to wonder what they would do in ours.

In my experience, people don't feel like they are "faking it" when they are enacting behaviors they value. On the contrary, most say it is at these times that they feel most true to themselves. Embodying traits you value can enhance positive mental health. When patients tell me, "I am who I am," my goal is to encourage them to understand that they are so much more.

Chapter 7

EVERYONE STUMBLES

G*rit*, the book by trailblazing psychologist Angela Duckworth, filled me with awe. As a longtime reader of Angela's papers in psychology journals, I expected her book to be engaging. Even so, I marveled that every page elegantly wove stories with studies to make a powerful point. Angela's book was smart. It was accessible. It was humble. It was funny. It was *her*.

After finishing *Grit*, I wrote Angela a gushy email about her book's effortless perfection. I mentioned I'd been struggling to write my own book and that I envied her skills as a natural storyteller. She emailed back and assured me that I was dead wrong. While she appreciated my kind words, she swiftly corrected my assumption about her apparent effortlessness.

"See attached for my early proposal for *Grit* . . . ," she emailed me back. "It's kinda terrible, see? and it looks nothing like the final version. trust me, girlfriend, I've been there!!!"

It's easy to assume that success comes easily to others, especially the people we admire. We endow our heroes with superhuman qualities and convince ourselves they are born winners and that failure is a foreign land they have never visited.

But everyone has spent some time in Failure—even if it's just a long weekend. I can't tell you how comforting it was to read Angela's barebones proposal.

A few weeks later Angela and I met up in New York, and she recounted in detail the blood, sweat, and tears (yes, she said she cried at her computer almost every day) that went into her book. The writing was not "effortless." Just the opposite. It was the hardest thing she had ever done. This new narrative helped me persevere. Angela, of course, knew this—after all, she studies persistence and motivation—and I have immense gratitude to her for pulling back the curtain on her own experience. The mountain was still steep, but I wasn't the only one who had stumbled on the way up.

We're often told that the best way to find inspiration is to dig deep within ourselves. I would argue the opposite: If you're trying to figure out what inspires you, look to others who share your values and whose life stories you admire and relate to. Olympic figure skater Sarah Hughes explains how role models are the ultimate source of inspiration: "Having role models is integral to staying motivated. It helps change the mindset from 'I'm tired. This is hard. Why bother?' to knowing it is possible to . . . be better." When Hughes was in college and training for the Olympics her role model was Tenley Albright, the first American woman to win gold in women's figure skating, who later went to medical school and became a surgeon.

"I was pursuing athletics and academics at the top level, something none of my peers were doing at the time—and something none of her peers had done either. Doing something no one else is doing is not easy. It can be very isolating. But thinking of how [Tenley] was able to excel in the rink and in school kept me motivated during the times I doubted it was possible to hand in an essay on the American Revolution in the morning and perform triple-triples in the afternoon."

Thinking about Tenley gave Hughes the courage and fortitude to persevere whenever she hit a rough patch. When your own resources are tapped out, outsourcing inspiration can help you stay strong.

Social Vaccines

The best role models don't live on pedestals, but instead are relatable. College women who took introductory classes in STEM (science, technology, engineering, and math) that were taught by female professors became more interested in the topics and more confident about their mastery of the material. Exposure to role models served as a "social vaccine," inoculating these young women against noxious stereotypes. During periods of personal transition, when self-doubt soars and uncertainty is a constant companion, social vaccines are particularly important.

In 2009 Michelle Obama visited the Elizabeth Garrett Anderson school in London, which served girls, many of whom came from low-income families. Mrs. Obama didn't give a generic speech about being a good student. Instead, she spoke to the girls about her own experiences growing up in a poor neighborhood in Chicago. She talked about how she had overcome challenges and made school a priority. Through hard work, she was admitted to Princeton and to Harvard Law School and ultimately landed a job at a prestigious law firm.

"[T]here was nothing in my story that would land me here," the former First Lady acknowledged. "I wasn't raised with wealth or resources or any social standing to speak of. If you want to know the reason why I'm standing here, it's because of education. I never cut class. I liked being smart. I loved being on time. I loved getting my work done. I thought being smart was cooler than anything in the world."

Mrs. Obama stayed in touch with the girls. When she returned to England two years later, she invited the same group to join her when she visited Oxford University. At Oxford she told them, "It's important that you know this—all of us believe that you belong here."

The students were inspired by Mrs. Obama's message. Following these interactions with her, their academic performance improved significantly. Economist Simon Burgess analyzed the girls' exam results and found dramatic improvement in their test scores that could not be explained by any other interventions at the school. It's possible that it was merely a coincidence or that the teachers had done something different with the class, but Mr. Burgess believes it was Mrs. Obama's relatability that boosted the students' belief in their own capacity to succeed.

Having a role model whose accomplishments seem within reach is a powerful source of motivation that can even benefit children as young as fifteen months. An MIT study found that babies who watched an adult struggle to reach a goal tried harder at their own difficult task (such as turning on a toy) than babies who saw an adult succeed effortlessly. Some parents feel the need to make everything look easy and not to get frustrated in front of their kids, but as this study shows, it might be beneficial to let them see you sweat a little. Children actively pick up skills from the adults around them. When they see them persisting in the face of a challenge, odds are they will, too.

When I attended the University of Pennsylvania to earn a master's degree in the field of applied science of positive psychology, I felt conflicted about not being home more with my kids, who were in grade school. Even when I was home, my attention was distracted by the sheer amount of research and reading. Sometimes the kids and I would do homework together. That was nice. Mostly I worried that my focus on my career was harming them in some way. A few years later I felt relieved after reading a study that concluded that the children of working mothers grow into happy adults. Although I'm still not feeling completely guilt free, I have shared this research with every working mom I know. From the moment a child is born, the guilt of potentially being an imperfect mother is born, too. Letting go of the pressure to be perfect frees you up to try to do your best.

It's laudable to have high standards, but unattainable ones erode vitality and undermine well-being. Students at Stanford University refer to the disconnect between an outward appearance of effortless perfection and the internal experience of struggle as the Duck Syndrome. (Picture a duck gliding calmly across the water, while beneath the surface, it frantically paddles its feet to stay afloat.) When all the other students seem to be gliding through school—getting good grades, thriving socially, going to all the cool parties, and landing enviable internships, it's easy to assume that you're the only one drowning in work and will never measure up.

The pressure to be perfect because everyone else around you seems to be perfect affects people of all ages. I've met many patients in their twenties through their fifties who feel tremendous pressure to exude confidence while struggling to keep it together. One mother in her early forties had a picture-perfect life—literally. Maria had moved from the suburbs to New York City. During our first meeting she showed me the Christmas card she'd sent a month earlier. It was a collage of "pinch me now" photos of her three kids frolicking in a turquoise ocean, her and her husband embracing in front of the Eiffel Tower, the family skiing, and the eldest child attending his graduation. Maria shared the card with me to convey the gulf between the facade she worked so hard to maintain and the emptiness she felt inside. She could talk to her new friends in the city about "surface problems"—what to do about a sick dog, the challenges of finding a good pediatrician, and synchronizing family schedules. Still, unrealistic expectations of perfection kept her from digging too deep without feeling embarrassed by her true feelings. She felt lonely in her marriage. Her son was struggling in his freshman year at college. She often wondered about what her life would have been like had she not given up on a career to raise her children. She was afraid if she voiced these concerns, she would be judged as being a bad wife and mother. The only people she felt comfortable confiding in were her sister and her psychiatrist.

Maria felt like a misfit on the island of flawless toys, convinced she was the only one in her social group having a hard time. Like many of us she overestimated the awesomeness of the lives of the people she saw around her and underestimated their struggles. A study, "Misery Has

More Company Than People Think," demonstrates how misperceptions of other people's picture-perfect lives can make us feel worse about our own. Although the message that "other people aren't as happy as you think they are" might not be the most comforting, it's worth taking into account, because to assume the lives of others are ideal can make you feel bad about your own.

Comparison Steals Joy

"Am I normal?" Deep down, this is the question many people would like to have an answer to. The need to belong is hardwired into all of us and doesn't vanish after high school. Feeling as if you're the only one who is out of sync with the rest of the crowd can amplify feelings of loneliness. Unhappiness, disconnection, and isolation will often follow. To paraphrase Theodore Roosevelt, comparison steals joy. It also is the thief of confidence, especially if your reference points are distorted. I told Maria about a study that found most first-year students believe that their classmates have far more active social lives and more friends than they actually do. Unsurprisingly, this insecurity is linked to lower levels of well-being and belonging. A grass-is-always-greener mentality can make our own lives look barren.

Maria's realization that her friends might be having a hard time, too, gave her the courage to open up to them. Over coffee Maria confided to one of them, who listened attentively and was supportive and thoughtful. The friend also revealed that when she had told everyone she was away visiting a sick family member a few months earlier, her story was a cover-up. She had really been hospitalized for depression. The feverish effort to maintain the appearance of effortless happiness, coupled with the belief that everyone else was effortlessly happy, had been exhausting and isolating for Maria. Discovering that she was not alone was a relief.

Feeling supported—and being able to support her friend—turned out to be central to her healing.

Anton Chekhov once wrote, "We see those who go to the market to buy food, who eat in the daytime and sleep at night, who prattle away, marry, grow old. . . . But we neither hear nor see those who suffer, and the terrible things in life are played out behind the scenes." This might explain the appeal of reality TV and websites like TMZ, which document the sordid personal lives of seemingly perfect celebrities. Seeing that everyone has problems is a reminder that you are not the only one dealing with them.

In the 1950s Leon Festinger formulated "social comparison theory," which was predicated on the idea that we determine our own social and personal worth based on how we compare with others. In the absence of an objective means of evaluation, we tend to constantly evaluate ourselves—our intelligence, attractiveness, wealth, possessions—against those around us. According to Festinger, people prefer to compare themselves with others who are similar to them. After all, what would be the point of a novice pianist comparing himself with Mozart?

There are upsides to social comparison. Students may feel more competent when they measure themselves against other students who didn't do as well on a biology test. Comparison can be a source of motivation—for example, a runner may want to emulate the performance of a fellow runner who beats her by a tenth of a second. Comparison can also make us feel more grateful for what we have and put disappointments and hardship into perspective. *Perhaps I don't have it so bad, after all* is a sentiment that often comes to mind when we think of others who are less fortunate.

That said, there are many downsides to social comparison. Research suggests that unhappy people make more frequent social comparisons than happy ones *and* it makes them feel worse. The tendency to seek social comparison is correlated with low self-esteem and depression.

Weighing ourselves against others may be in our DNA, but the context has recently changed dramatically with the rise of social media. When we see people in social contexts, laughing and having a great time, many of us assume this is the regular tenor of their lives. Pulling back the curtain, witnessing their challenges, and recognizing their humanity can provide some much-needed perspective.

Rather than making comparisons with people who are in the same position as we are, we now have a global landscape that is packed with endless visual streams of perfect faces, butts, parties, romances, and vacations. Women who frequently measure themselves against other women are especially vulnerable to feelings of inadequacy about their appearance. Just a few minutes of looking at images of someone you think is better looking than you are can bring you down.

When I find myself doing this, I remind myself that these images are typically manufactured. When images are viewed through this perspective, they have less of a negative effect. One study found that exposing the performative side of Instagram by juxtaposing photoshopped pictures with the real ones can help women feel better about their bodies.

Fixating on ideals depicted throughout social media—and in all aspects of our lives—distracts us from actually doing what's possible. Role models that are supermodels or unrelatable superstars may only increase insecurity and pour the proverbial gasoline on the fires of self-doubt. When our attention is consumed by the appearance and pursuit of perfection, we close ourselves off from opportunities, unwilling to even take a chance for fear of failure. Social connections, curiosity, and perseverance wither in the face of unrealistic expectations. The antidote is to find realistic role models who inspire you and to be a realistic role model for others.

Chapter 8

BETTER DAYS

Pathogenesis—the treatment of disease—is not
the same as salutogenesis—the creation of health

Think for a moment about all the decisions we make every day, from what to wear to what to eat to whose email to return. The urgency and frequency of decisions that demand our attention and time seem to increase with every hour. We are bombarded with endless choices to be made and actions to be taken, from the simple "cream or milk?" to more significant ones like, "Should I change jobs?"

Choices are unavoidable, and to make things harder, everyday stress can derail our ability to make sound decisions. When we're anxious or upset, we're less likely to believe we can tackle a challenge or try something new. Confidence in our ability to prevail plummets, and self-doubt sets in.

The results of one study indicated that when people are made to feel anxious they are more likely to take bad advice, even when it's obviously bad, because they don't trust their own ability to make a good decision. We become susceptible to people who might not have our best interests at heart, whether a real estate broker ("This apartment is a great deal—you should snap it up today"), a fitness instructor ("Work through the pain"), or frenemy ("Nobody will notice if you don't show up at the office party").

It is impossible to rid ourselves of most of the irritations and anxieties that pop up in daily life. What *is* possible is to stay fortified in the face of them. Unfortunately, our culture offers a false narrative about what actually makes us feel better.

I had a patient who would spend hours poring over clothes and shoes on expensive retail websites. Entire Saturday afternoons would pass as she added to a cart the latest looks, which she would never purchase. Another patient would spend hours studying real estate listings. "If only I lived there," she dreamed. It was "privilege porn."

We desire things that are shiny and new. As Carrie Bradshaw reminded us in every episode of *Sex and the City* and Instagram influencers whisper in our ears today, everything will be okay if you have the right handbag or that cute pair of shoes. Research tells a different story. The accumulation of wealth and possessions is not a reliable way to sustainably promote enduring happiness. The thrill of getting that new bag quickly dissipates. The tendency to get accustomed to having nice things is known as "hedonic adaptation." We quickly adapt to repeated acquisition, which explains why you were over the moon when you bought that bag but now that a celebrity has the new season's version, you want that one.

"Passive leisure" is another uplift imposter. Surfing the internet, checking social media, and watching TV seem like energy restorers but often leave us feeling even more drained. Low-effort activities that ask so little and require minimal physical, intellectual, or social involvement do not build the same lasting positive resources as more engaged activities. They appeal to our instinct for idleness but do not sustainably revitalize us.

Cultivate Uplifts

Everyday well-being resides not just in our heads but in the actions we take, the connections we make, and how we participate.

When patients with mental illness are asked about what they think scientists should study, better treatment is a top priority. They care about discovery of more effective medication and gaining greater understanding of the underlying illness. But that's not all. They're also concerned about having a better quality of their daily lives. They care about possessing the ability to function efficiently and effectively both emotionally and physically. When asked what wellness means, a respondent in a survey wrote, "It . . . means [being] well enough to enjoy activities, well enough to feel joy and hope."

While studying the impact of everyday annoyances on well-being, psychologist Anita DeLongis and her colleagues at UC Berkeley recognized how even minor positive events—an easy commute home, a laugh with a friend, receiving praise from a teacher or boss—prevented or attenuated the experience of stress and promoted health. They showed how uplifts counteract the negative impact of the daily grind by serving as "breathers" from stressful encounters, as "sustainers" of coping activity, and as "restorers" of depleted resources. In the absence of uplifts, "microstressors" can snowball into a potent and harmful form of chronic stress that could leave a person vulnerable to depression and other types of mental illness.

Today there is increasing evidence that positive everyday experiences and activities that engage, connect, and fortify us are critical sources of vitality. By generating positive emotions, such activities have been shown to strengthen resistance to everyday challenges and also to promote recovery from them when they do occur. I think of positive activities as emotional armor.

In an ideal world, we could wave a magic wand and get rid of all the annoyances and irritations that interrupt our days and keep us awake at night. But in the real world, that's just not possible. Nor does simply "thinking positive" provide an answer. If a pen leaks all over your blouse, sure, you can tell yourself that you didn't really like the blouse anyway or that the spill is an "opportunity" to learn how to remove ink stains. But it's still really bothersome. It's a messy situation. There's ink all over your hands. Cleaning up is going to take a while. It's going to be expensive to have the blouse cleaned properly, and it will probably never be the same.

Positive emotions counter the tendency to isolate and the inclination to be rigid. They intercept the telescoping effect of a bad mood and instead promote flexibility in thinking and problem-solving and give rise to positive actions. Positive emotions also motivate people to become more interested in participating in social and physical activities.

Importantly, positive emotions can occur amid daily challenges and coexist alongside frustration. We tend to see our days in terms of absolutes, but life rarely presents an either/or situation. Sleeping through your alarm might make it seem as if the rest of the day is doomed, but believing that to be the case can become a self-fulfilling prophecy. If you are already having a tough morning, it is far more likely that you will be more aggravated than usual when you find out the delivery of your lunch order is going to take an extra twenty minutes. An uplift can keep such a minor annoyance from spiraling you into a bad day. When it rains it doesn't have to pour.

This advice isn't rainbows-and-unicorns pop psychology. Scientific evidence has demonstrated that uplifts can make a person more Teflon and less Velcro, both psychologically *and* physically. Positive emotions can "undo" such cardiovascular effects as increased heart rate and blood pressure from the daily grind. They can even make you less likely to catch a cold and suffer inflammation.

Some hassles produce more negative emotions than others. Arguments with loved ones is high on the list. But even those that aren't personal can feel like targeted attacks. One day my patient Mark, a banker in his forties, came to his session fuming.

"What's going on?" I asked.

"My flight to Chicago was canceled. Weather or some other BS. This is absurd. Why me?"

Obviously, a canceled flight isn't a personal matter. But events that are beyond your control or disrupt your routine do have the ability to make you feel particularly powerless.

The more negatively you respond to a given situation, the more stressful you experience it to be, and the more lasting impact it will have on your mood. If you are in a negative place, you are more likely to interpret the next potential hassle as threatening. You are more likely to take a setback personally ("This is all my fault"), assume it's permanent ("This will keep happening no matter what I do"), and view it as pervasive ("This messes everything up"). When you're in a more positive frame of mind, you are more likely to interpret a bother as a challenge. Simply put, your energy and mood influence whether or not a potential hassle becomes an actual one.

Upward Spirals of Growth

An uplift can help reduce the negativity experienced in the wake of a frustrating interaction or a pointless meeting. Uplifts are more than feel-good-in-the-moment positive experiences. Although they may be short lived, the coordinated changes an uplift can produce in people's thoughts, actions, and physiological responses are lasting. They build social, intellectual, and physical resources, and create what researcher Barbara Fredrickson calls an "upward spiral" of growth. She wrote, "Negative emotions narrow one's momentary thought–action repertoire by preparing one to behave in a specific way." Fear triggers an urge to escape. Anger triggers an urge to attack. Disgust triggers an urge to distance ourselves from whatever offends us. In contrast, positive emotions widen possibilities for action. Joy, contentment, and interest expand the range of cognitions and behaviors.

Expanded thinking and attention encourage creative solutions to problems and the urge to approach, act, play, and explore. When you're in a good mood, you're more curious, more open to learning something new. You're more patient with a coworker, more willing to lend someone a hand. You're even more likely to notice a bird flying overhead. These brief "lived experiences" build on one another, engender vitality, and cultivate resilience. Fredrickson likens positive emotions to nutrients that we need in order to grow and be healthy.

Sources of uplifts are rarely self-focused. They typically occur in connection with others and involve doing rather than thinking. They are

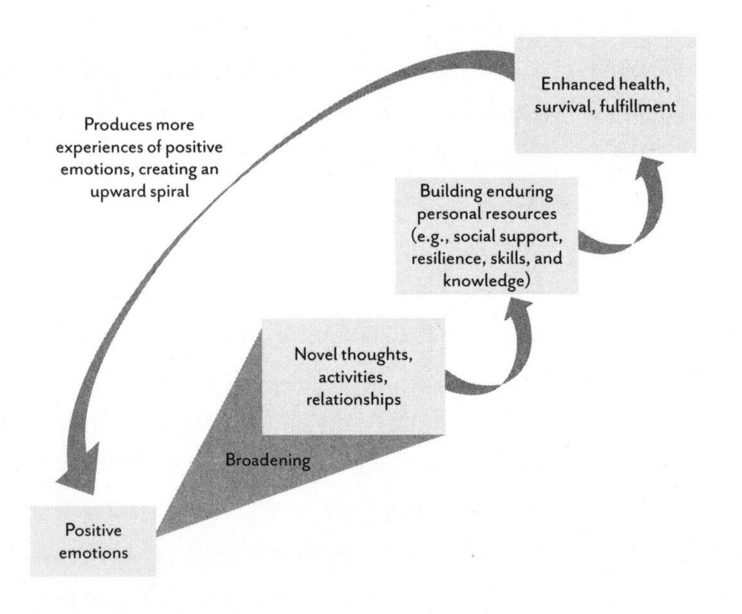

The broaden-and-build theory of positive emotions.
From Barbara L. Frederickson, "Positive Emotions Broaden and Build." In Vol. 47 of Advances in Experimental Social Psychology, *edited by Patricia Devine and Ashby Plant (Burlington, MA: Academic Press, 2013), 53.*

often action-oriented and other-oriented. Thinking happy thoughts doesn't qualify as an uplift. Nor does passively scrolling through Instagram or mindlessly flipping through channels on the TV. These are uplift imposters. For an uplift to provide a genuine boost, it requires one's full attention and involves at least a micromoment of connection or engagement with something or someone. It might be learning something new or doing something as simple as holding a door open for someone. It might be working on a project or reading a few pages in the novel on your bedside table. It might be chatting with the taxi driver or giving directions to a stranger. It might be cooking dinner for your family. These small moments of grace or goodness are the essence of everyday vitality.

Generating uplifts in your daily life has other benefits. Research shows that it can even help to boost self-control. Willpower typically pulls a disappearing act in the face of daily frustrations. On tough days, you are more likely to smoke that occasional cigarette or have an extra drink. The more irritation and frustration you experience throughout a given day, the more likely you are to reach for a doughnut or a bag of potato chips. It is especially difficult to resist the allure of high-fat and high-sugar snacks when you're annoyed or upset. Willpower comes and goes. It might help you bypass the candy aisle at the grocery store but be long gone by the time you reach the gauntlet of tempting treats at the checkout counter.

Instead of trying in vain to summon more willpower, research offers an alternative path to persistence. Elevating levels of positive emotions in daily life is a more reliable way to cultivate self-control. A study found that people who engage in prosocial actions like donating money to charity or helping another person exhibited greater self-control than those who didn't. As researcher Kurt Gray suggests, "Perhaps the best way to resist the doughnuts at work is to donate your change in the morning to a worthy cause."

Some very smart people believe that white-knuckled willpower is the only key to success, and they inevitably cite Walter Mischel's famous marshmallow study. In that experiment, young children were given a choice—they could either eat one marshmallow as soon as they felt like it, or wait for the experimenter to return to the study room, at which point they would get two marshmallows. Some gobbled up the marshmallow without hesitation, but most did their best to resist temptation. The follow-up study conducted forty years later revealed eye-opening data about the adult lives of the marshmallow-munching kids. Those who were able to delay gratification went on to attain more success in high school, receive higher SAT scores, earn better grades in college, land better jobs, cultivate better relationships, and were even in better physical shape than those who couldn't wait.

As a result, policy makers, educators, psychologists, and parents have all jumped on the willpower bandwagon. They are convinced the key to greater self-control is learning how to overcome temptation by changing how you think about whatever lures you. Author Pamela Druckerman summed up this approach in an op-ed in *The New York Times*: "Don't eye the basket of bread; just take it off the table. . . . When a waiter offers chocolate mousse, imagine that a cockroach has just crawled across it."

Although these strategies can be effective, they are not foolproof. (Also, who wants to think about a cockroach racing on top of chocolate mousse under any circumstances?) Genetics, environment, stress, and fatigue, among other factors, play a role in the ability to resist temptation. It takes a great deal of energy and effort to combat "hot" emotional impulses with cognitive cool thinking. Plus, suppressing emotions is stressful and takes a toll physically and emotionally.

With all the focus on devising strategies to increase willpower, we may have been neglecting one right under our noses—the benefit of uplifts. In more recent marshmallow-related news, children were more willing to wait for the treat when paired with another child. Participants in both Kenya and Germany found strength when they knew it benefited someone else.

Chapter 9

MORE LIFE

Uplifts are easily generated because they don't require either money or extraordinary circumstances. Still, they are not effortless to produce. They have to be created, noticed, and prioritized. When study researchers asked nurses, assistants, and receptionists at an outpatient family practice clinic to pay explicit attention to good things that happened in their day, in just three weeks they all reported feeling significantly better. All they were asked to do was write down what had gone well that day and why. This simple practice reduced stress in the moment and also later in the day. It also minimized mental and physical complaints like headaches and muscle tension and enhanced their ability to detach from work in the evening.

"What's great about this exercise, is the power it gives to each of us on a daily basis," the authors of the study noted. "Before turning on the radio or getting on a call during your homeward commute, take a moment to reflect on the good things that happened at work. Doing so can help you capitalize on the small, naturally occurring flow of daily positive events—a ubiquitous but too-often-ignored source of strength and well-being."

Decades of stress research have focused on removing stressors, reducing pressure, and decreasing mental strain, but it's equally worth exploring the potential of everyday positive experiences. We often separate "work" and "life" as if we're waiting for work to end so that we can get back to living. But given that we spend more than a third of our lives on the job, we're leaving a lot of potential uplift on the table. In Studs Terkel's oral history *Working*, he noted the importance of integrating the two. "[Work] is about a search, too, for daily meaning as well as daily bread, for recognition as well as cash, for astonishment rather than torpor; in short, for a sort of life rather than a Monday through Friday sort of dying."

Can I Eat It? Can I Have Sex With It? Will It Kill Me?

Highlighting accomplishments, sharing positive experiences, celebrating success, and encouraging growth are all ways that a day, even a day full of irritations, can feel less like dying and more like living. But focusing on positive moments and experiences does not come naturally for most of us. This probably has an evolutionary basis, as ancestors highly attuned to threats were more likely to survive. When those ancestors encountered something unfamiliar, they placed it in one of two categories: friend or foe. Should it be approached or avoided? Deciding whether something was a food source, of reproductive value, or a predator made the difference between life and death. Can I eat it? Can I have sex with it? Will it kill me? were the three fundamental questions that had to be solved by primitive brains.

Fear serves an important function. When a gazelle is chased by a lion, her body goes into survival mode: stress hormones flood her bloodstream, her heart goes into overdrive, extra blood is shunted to her muscles to facilitate running, her awareness intensifies, and her vision sharpens as her body devotes all its energy to escaping the beast.

It is pretty much the same with humans. Think of swerving out of the way of an oncoming truck—you turn the steering wheel before you even register you are afraid. Your reaction is instinctive.

If the gazelle outruns the lion and survives, she returns to the watering hole with her fellow gazelles and resumes her routine hanging out and eating grass. Her stress hormones return to normal, along with her blood pressure and heart rate. The gazelle doesn't dwell on how alarming it was to have been chased by a lion, nor does the fear of another attack consume her.

This is where humans and gazelles differ. It isn't as easy for us to move on. Inconveniences, interruptions, setbacks, and challenges are all perceived as threats that command our attention and energy throughout a busy day. This mindset explains why, for example, an employee can receive an annual review that is 99 percent complimentary and still focus exclusively on the 1 percent that was critical.

Frustration, fear, anger, and disappointment dominate our emotions, narrowing our focus and keeping us on guard, ready for battle. Although these emotions do contribute to our survival, they also interfere with our quality of life.

THE ZEIGARNIK EFFECT

Negativity bias explains so much of the tenor of our internal thoughts. We beat ourselves up for not finishing a project we hoped to get done by the end of the day without giving ourselves credit for all we *did* accomplish. The phenomenon of unfinished business sticking with us is so prevalent it even has a name: the Zeigarnik effect.

While traveling in Vienna, psychologist Bluma Zeigarnik was struck that waiters in a café easily remembered orders that were in progress but would forget them once the bills had been settled. She constructed an experiment that gave participants math problems, puzzles, and other tasks to perform. During some of the tasks, participants were interrupted. Afterward, they were asked which of the activities they remembered. They were twice as likely to recall the tasks that they hadn't finished. The Zeigarnik effect helps explain why you can complete a project at work, help a friend make a big decision, cook dinner, and return fifty-two emails, and yet, when you get into bed, all you can think about is the one memo you forgot to send.

The Zeigarnik effect can be helpful when it motivates us, such as when that nagging feeling gets us up early to work on a project. But it can also unleash worry and anxiety when we have too much on our plate. In times like these, we try to offset the anxiety by turning to "productivity porn"—the books, apps, and seminars that promise to help people learn to work more efficiently. Productivity porn offers a smorgasbord of tips and tricks that help you squeeze more into your day. But there's a cheaper, simpler way to get more done: cultivate more uplifts.

Researchers who conducted one study asked participants at the beginning of each day about the goals they hoped to make progress on that day. At the end of each day they rated their progress toward achieving these goals. On days when participants experienced a high number of annoyances (such as a dead cell phone battery, or an endless traffic jam) and relatively few uplifts, they made less progress. But on days they reported experiencing a relatively high number of uplifts (doing something with friends, watching the sun go down) the annoyances were buffered and did not interfere with their goals. Put simply, "time-wasting activities" might actually be time well spent. It turns out that a walk around the block or taking a moment to look out the window, enhance our ability to achieve our goals. A few minutes of friendly conversation can actually boost problem-solving skills and the ability to resist distraction.

Vitality Is a Verb

Just because we are drawn to the negative does not mean we must succumb to those feelings. By intentionally noticing and generating uplifts and engaging in health-promoting and vitality-enhancing activities, it is possible to overcome the gravitational pull toward obsessing about what can, did, or will go wrong. Compared with people who focus on hassles in their daily lives, when we focus on appreciating someone or something outside of ourselves, we are more likely to create the conditions for vitality. We are also more likely to exercise and sleep better. Perhaps, most important, we are more likely to help someone else.

Not long ago, in our last session before the new year, a young patient who occupied that gray zone of not qualifying for a formal diagnosis of depression but still feeling overwhelmed and exhausted by her life told me, "So, this year I am going to be like Drake."

My kids were then too young, and I was too old to be a fan of the Canadian superstar, and my bewilderment must have registered, because she then explained, "*More Life*—it's the name of Drake's album. That's my goal." This statement captured what she and so many other people I meet long for.

Another of my patients, Amelia, had always been an overachiever. During college exams she avoided the library, not wanting to be distracted by others, and instead would hole up in her room and go to the dining hall during off hours. When she entered the work world, she continued this isolating behavior, which she thought maximized her productivity. She didn't have time for "niceties" at the office—no chitchat with colleagues, no lingering on lunch breaks. She believed anything that interfered with getting work done was not time well spent. Amelia thought of herself as a racehorse with blinders on.

When other people showed her kindness, she assumed they probably needed something. If a project was going well, she would envision the many ways it could go wrong. If a guy expressed interest in her, she concluded there must be something wrong with him. She was like an inspector with a flashlight entering a dark room. There might have been artwork on the walls or flowers on a table, but she only directed her flashlight in the corners, looking for cockroaches and cobwebs.

In some ways Amelia reminded me of my younger self. For years I focused on the cockroaches and cobwebs. Like my patient, I found something comforting about swimming in a sea of worry and work. It prevented me from feeling vulnerable; it also prevented me from having a vital life.

In therapy I learned it was possible to retrain my brain. My flashlight didn't only have to search for threats. I could pay explicit attention to what had gone well and even give myself credit for it. I could feel pride in my accomplishments, which helped offset the worry about what I hadn't finished. I also realized that generating uplifts buffered hassles and helped me be more productive. Shifting my attention did—and still does not—come naturally to me, and I work on it almost daily. Amelia does now, too. We both understand that vitality doesn't just happen, it has to be created with intent.

Chapter 10

TAKE ACTION

Happiness is when what you think, what you say, and what you do are in harmony.

MAHATMA GANDHI

Sarah, a twenty-three-year-old recent college graduate, had seen a psychiatrist during college for mild symptoms of depression. With therapy and medication, she felt less depressed, less pessimistic, and less worried. Still, her brighter mood didn't translate into her taking any concrete action with respect to her future. After graduation, she moved back in with her parents and spent her days on the sofa in front of the television watching talk shows, scrolling through social media, or running errands. She daydreamed about moving out, but she first needed a job. She had fantasies of finding a boyfriend but refused to be set up by friends or to join a dating website. She was stymied by fear of rejection—"It's why I have such a hard time putting myself out there"—and had spoken extensively with her college therapist about her fears. Understanding and insight, however, did not help her overcome those fears or motivate her to make a change, nor did thinking or talking about how great it would be to have a nice apartment and a handsome boyfriend.

Insight without follow-through left Sarah spinning her wheels. In our sessions I steered her to focus on proactive steps she could take. She had mentioned enjoying doing community service work when she was in high school, so I asked her to come up with a list of organizations where she could volunteer once a week. She chose a nursing home a few blocks from her house and signed up to work there every Wednesday. She was surprised by how much she enjoyed spending time at the home, playing cards and reading the newspaper to the elderly. She liked feeling useful. The positive experience in the nursing home emboldened her to take concrete steps in other areas of her life as well. She signed up for spinning classes twice a week with a friend. She also put together a résumé and submitted it to a health-care startup. She created an online profile on bumble.com and went on two dates within ten days. The dates were "not terrible," and she even started seeing one of the men more regularly.

Working at the nursing home led to a shift in Sarah's perspective. She realized how taking positive actions, and not just thinking more positively, made a difference. The link between action and vitality began to become clear for her.

In my practice I spent years trying to change how people think through insight and self-reflection. I believed that if I could help them become more open-minded or optimistic, or less judgmental they would be able to initiate the change they longed for. But simply talking through problems doesn't necessarily result in behavior change.

Behavioral Activation Therapy

A treatment plan that emphasizes action is known as behavioral activation (BA) therapy. It's an "outside-in" approach that emphasizes how a person acts in the real world, as opposed to an "inside-out" approach that focuses on a patient's thoughts, which is an approach that characterizes most treatments. Behavioral activation is predicated on the idea that engaging in rewarding and satisfying activities can alter how a person feels and counteract depressive pastimes like rumination and avoidance. A study in *The Lancet* showed that BA therapy is as effective at treating depression as the gold standard "inside-out" treatment, cognitive behavioral therapy, which primarily focuses on reframing problematic thoughts and disrupting negative thinking patterns. Another study of 241 depressed patients revealed BA therapy to be as effective as antidepressant medication.

It is common knowledge that how we feel affects our behavior. If you are in a bad mood, you are much more likely to lose your temper or snap at someone. Less obvious is that *what you do also shapes how you feel.* When people seek out engaging, connecting, and learning experiences in their daily lives, they feel stronger and more vital. Instead of avoiding difficult situations or passively allowing in-the-moment feelings to take precedence, BA encourages planning and specific and decisive actions. A study of depressed university students showed that just five sessions of BA therapy improved their symptoms and also affected their brain circuitry. Brain scans revealed that BA reactivated areas of the brain that become sluggish when depressed.

A brighter mood and thinking more positively only took Sarah so far. The key was bringing her into contact with situations she had previously found to be uncomfortable and creating experiences that provided a sense of accomplishment and agency. After all, the best way to increase competence is to actually accomplish what you put your mind to. Doing something that feels meaningful or that matters to you, and not just thinking about it, is uplifting.

Imagine having an hour free at the end of the day and deciding to sit down to stream an episode of your favorite television show. When the episode ends, it's time to go to bed, but before you can get out from under your cozy blanket, the next episode begins to stream automatically. The choice to watch the first episode was on you, but the next three were on Autoplay. By the time you finally do get to sleep, it's late. Your schedule is thrown off, and you sacrifice your well-being for a decision you didn't really make. It's easy to drift, slip, and slide into behavior that moves you away from your original plan. Think about the situation of a couple who decide to live together because one partner's lease runs out. Sure, there are financial advantages, but if that's the main reason for cohabiting, they're on relationship Autoplay. Couples who make a deliberate decision to live together compared with those who shrug their shoulders and say, "It just kinda happened," experience a marriage of greater quality later on. Making conscious choices protects relationships.

A passive existence is not a vital one. Living a life on Autoplay, shrinking from challenges, isolating oneself, dwelling on the past, and burrowing internally remove us from the experiences that vitalize us. Indeed, vitality is a reflection of how we actually choose to *live*.

Doing, Not Dreaming

Talking about your problems, policing your thoughts, and thinking positively doesn't automatically lead to constructive change. New York University professor of psychology Gabriele Oettingen has found that people who spend time picturing how good it will feel to reach a goal without taking any concrete actions toward it are *less* likely to achieve it. In a study of obese women who enrolled in a weight-loss plan, Oettingen found that the women who had positive fantasies about their weight loss—such as showing their new body off to a friend who had not seen them in a year, or supposing that it would be easy to resist a leftover box of doughnuts—were less likely to lose weight than those who were realistic about the challenges they faced.

Oettingen found a similar pattern across multiple domains, including quitting smoking, starting a relationship, doing well on an exam, and getting a job. Fantasizing about being successful without actually pursuing it also undermined motivation. Dreaming turned out to be devitalizing. In fact, participants in a study who were asked to generate positive fantasies about the week ahead felt *less* energized and later reported poor accomplishment and lower mastery of everyday challenges. Moreover, they were less likely to put in the effort and persist when setbacks occurred. Oettingen theorizes that mentally attaining what you want obscures the actual need to apply the effort to make it happen.

Mental Contrasting

Instead of fantasizing, try mental contrasting, which combines being optimistic with being realistic. Mental contrasting means imagining a positive outcome while recognizing the potential obstacles involved and planning actions to overcome them. Based on her research about mental contrasting, Oettingen recommends setting what she calls WOOP goals to close the gap between one's present reality and desired future. The four steps are:

1. **W**ish: What is something you wish for? Make a **wish**.
   Imagine something meaningful and important to you that can be attained within a specific time frame. Put the goal into words.
   Examples: "I want to do well on my math test." "I want to feel more gratitude while living my life."
2. **O**utcome: What would be the **outcome** of fulfilling that wish?
   How would you feel? Imagine feeling this way and put it into words.
   Examples: "I would feel deeply engaged in my work." "I would feel proud of myself." "I would feel tremendous relief."
3. **O**bstacle: What is the number one **obstacle** that might stand in your way?
   Consider what can hold you back from achieving what you wish for. Say it to yourself or put it down in words.

Examples: "I have a hard time saying no." "I get distracted by social media." "I am always exhausted." "I am a procrastinator."

4. **Plan:** What is your **plan** to overcome this obstacle?
   What is an executive action you can take to tackle this obstacle? Make what is known as implementation intention (aka: an action plan) to confront the obstacle when it arises.
   Example: "If someone offers me a drink, I will say, 'No, Thank you.'" "If I get distracted by my phone when I am with my family in the evening, I will leave it on my desk." "If I feel like eating junk food, I will go for a walk around the block."

Personalizing challenges and determining how to overcome them increases agency and promotes commitment. Mental contrasting has been found to be an effective technique to increase physical activity in stressed-out college kids. Students improved their grades and time management using WOOP. Nurses who performed a daily WOOP exercise to reduce work stress reported reduced psychological and physical symptoms and increased vigor and engagement after three weeks. Importantly, all of the nurses generated their own stress-reducing wishes and identified what the outcome of achieving them would mean.

The strategy of mental contrasting can also be leveraged to reduce unhealthy behavior in a relationship. Moments of insecurity are inevitable for a couple, and even the most confident individuals with committed partners worry about rejection. This isn't neediness—it's human. Love is always accompanied by the risk of being stripped of that love. When one of my patients went to Las Vegas for a bachelor party, his girlfriend of three years called and texted him many times throughout the weekend. Franklin responded to most but not all of her messages. When he returned home, she was distant and aloof. Although Franklin had stayed up late gambling, he hadn't engaged in any behavior that might have been disrespectful or upsetting to his girlfriend. He brought her a present—a silver bracelet with a charm of the Eiffel Tower, commemorating the fact they'd met in Paris. He thought she would be touched, but instead the gift only fueled her feelings of insecurity and perpetuated the coldness.

Franklin's girlfriend rebuffed any attempt he made to express affection over the following few days. She later explained in a couples session how embarrassed she had been at attempting to contact him so often while he was away. She was convinced that her neediness was obvious and that the only reason he gave her the present was to appease her. To protect herself, she pushed him further away.

Whenever you're tempted to look through your partner's email, or check her phone log, or engage in other insecurity-driven actions, mental contrasting can help to overcome that urge. Remember the four steps to setting WOOP goals: If your *wish* is to stop checking in on your partner because the *outcome* would be more trust, but the *obstacle* is insecurity, then, a potential *plan* might be: "If I feel needy or jealous, then I will call my best friend." Proactively orienting yourself away from the obstacle and toward an actionable alternative will help prevent you from engaging in self-sabotaging behavior.

The Tumbleweed Factor

Jack's job required him to spend a lot of time on the road, and even when he was home, his schedule seldom gave him free time. He had two young daughters whom he adored. He longed to spend more time with them, but he often had to rush off to breakfast meetings in the morning and return when the girls were already asleep. On weekends he would accompany them to birthday parties and playdates, but he wanted more cozy father-daughter time. Every Sunday night he'd promise himself that the upcoming week would be different.

A significant part of Jack's day was spent communicating with his company's West Coast office, which meant his days began relatively quietly and then would rev up as the California office got busy. It wasn't just the scheduled events on the calendar that kept him occupied, but any number of pressing matters would pop up, demanding his immediate attention. Before he knew it, it was Friday night, and once again he had barely seen his daughters.

Jack's best intentions were frustrated by what I call the "tumbleweed factor," a sense of being helplessly rolled around about by the wind. My patient recognized how all too often he'd passively accept agendas and suggestions presented by others. Although a great deal of his life was dictated by forces beyond his control, it was essential that he begin to tease out what he actually did have a say in. We talked through a number of options. Seeing the girls before bed was not always a realistic possibility, so he came up with a more feasible plan: if he wasn't on a business trip, Jack would walk them to school in the morning, every morning. No exceptions. Just he and the girls. If someone proposed a breakfast meeting, he knew exactly how to respond: "Sorry, breakfast won't work. Can we set something up for later?" He blocked the time on his iPhone calendar so he wouldn't inadvertently add a meeting. Jack also informed his assistant that 8:30 to 9:00 a.m. was permanently unavailable; that time became sacred.

Rain or shine, hand in hand, Jack and his daughters walked (or skipped) to school. These morning walks became the highlight of his day and had a halo effect that made the rest of the day seem sunnier. His workload never changed, but he felt less stressed by it. The morning routine vitalized him. When actions align with intentions, you fortify yourself against the chaos of everyday life. Jack told me he was no longer haunted by the "could have, should have, would have" thoughts that used to keep him awake at night.

"You know why?" he asked. "Because I *do*."

At the heart of WOOP-goal strategy is anticipating the real-life hurdles that might undermine your true desires. To plan and prioritize spares you from having to make an in-the-moment impulse-driven decision. Whatever is happening to you *right now* loses its grip. Hot emotions like anger and worry are no longer all-powerful. Having a go-to action better enables you to override counterproductive inclinations and prevents you from being overrun by fleeting thoughts and emotionally driven wants.

Decide to Decide

People so often find themselves tongue-tied in the moment, unable to come up with the perfect retort. Only later, when they're headed home, does the ideal response occur to them. The French have an expression for this frustrating experience—*esprit de l'escalier*, which translates to "wit of the staircase." Patients often talk about what they wish they had said at a particular moment. They relive conversations in their minds and wonder, *Why didn't I think of XYZ?* They wish they could go back and render the other person speechless with their devastatingly witty reply or amusing zinger.

When it comes to lingering regret, though, failing to enact one's values, especially in relation to others, has an even more enduring impact. Several years ago I had an onstage conversation with Buddhist teacher Dzogchen Ponlop Rinpoche, during which he shared a moving parable that captures the staying power and consequences of passive choices.

> *Two monks were walking together in the woods. At one point, they came to a river with a strong current. As the monks were preparing to cross the river, they noticed a frail woman also attempting to cross. The woman asked if they could help her cross to the other side. The two monks glanced at one another because they had taken vows not to touch a woman. Then, without a word, one of the monks picked up*

> *the woman, carried her across the river, placed her gently on the other side, and carried on his journey.*
>
> *The other monk couldn't believe what had just happened. After rejoining his companion, he was speechless, and an hour passed without a word between them. Finally, the baffled monk could not contain himself any longer, and blurted out "As monks, we are not permitted to touch a woman, how could you carry that woman on your shoulders?" The other monk looked at him and replied, "Brother, I set her down on the other side of the river, why are you still carrying her?"*

Passively going about one's business can be an on-ramp to regret and rumination. Like the monk who didn't help the woman, we "carry" inaction with us. In his 2013 commencement address to the graduating class of Syracuse University, the writer George Saunders spoke about what he wished he had done differently in his life. He told a revealing story of a new girl who joined his class in the seventh grade. She was shy and awkward, and some of his classmates made fun of her. She had no one to sit with in the cafeteria or to play with during recess. Her family moved away a few months later, and he never saw her again.

"Why, forty-two years later, am I still thinking about it?" Saunders asked. "Relative to most of the other kids, I was actually pretty nice to her. I never said an unkind word to her. In fact, I sometimes even (mildly) defended her. But still. It bothers me. So here's something I know to be true, although it's a little corny, and I don't quite know what to do with it: what I regret most in my life are failures of kindness."

After all these years Saunders is still "carrying" that girl from seventh grade. The most persistent regrets of so many people are deeds they wish they'd done, not deeds they did but wish they hadn't. Those deeds can be large or small. Compare the regret of not calling a grandparent for a month right before that grandparent dies with the effort of making that call. One of my patients regrets skipping the wedding of a close friend because (excuse alert) the patient was working hard and chose not to make the effort to attend. Decades later—even though the couple who got married is now divorced—the patient carries the pain of this decision with her. She told me the silver lining of her experience is that she uses this story to teach her kids about the importance of showing up for friends.

If you want to optimize your vitality, put guardrails in place so you don't get to the end of the week (or month . . . or year . . . or decade) and regret your choices. Of course preventing or anticipating every gust of wind is not possible, but establishing a safety mechanism so those winds don't blow you off course *is* possible. With a structure in place, intentions are more likely to translate into actions. To live a more deliberate and vital life, don't slide into your decisions. Actively decide.

PART THREE

Connect with Others

Chapter 11

SKIP THE CASSEROLES

In the wake of a major life event—for example, the loss of a loved one or a serious illness—we often receive an outpouring of social support. Coworkers offer assistance. Friends and family step in with an abundance of affection and generosity. Neighbors show up with casseroles.

Because such events are typically visible to others, a sufferer's needs become obvious. If someone's house burns down, he clearly requires a place to stay. If someone's parent dies, we send her condolence notes and flowers. The ways to give and receive support are part of customary social behavior.

There is little comparable help offered for daily stresses, which may be part of the reason why these hassles take such a toll. Nobody drops off a meatloaf because they heard you had a rough commute. Hugs are not usually available because you forgot your umbrella and got soaked in the rain. (Who wants to hug someone who's wet?) Friends don't call or text you because they are concerned you couldn't find a parking spot and ended up late for your doctor's appointment.

The need for social support and the benefit of positive interactions are less obvious for buffering everyday stress. I meet many people who work hard at doing well at their jobs and invest time and energy into paying attention to deadlines and completing projects. They are conscientious about arriving to meetings on time and returning emails—so much so, they think nothing of leaving their significant other sitting in a restaurant, waiting, while they return an email in the restroom. After a long day being polite to coworkers, they snap at their partner. They make time to return work calls, while ignoring one from a friend. What is urgent often eclipses what's important. Our relations with others get sidelined as the demands and challenges of the daily grind grab our attention, drain our energy, and hijack our time.

When I went to medical school, there was always more studying to do, always more to learn. Becoming a doctor captured my total focus. I skipped birthdays, baptisms, and weddings. I rarely made time to see friends or family. One year I even forgot Father's Day. At the same time I was almost pathologically polite to my patients, teachers, and fellow students. I'd offer an apology to a stranger who stepped on my toe but was emotionally absent from those to whom I was close. My assumption was that my friends and family would understand and forgive my lack of effort in maintaining our relationships. So when I found out that I hadn't been invited to a friend's small birthday dinner, I got upset—at her. I ranted to my sister that this friend had been cruel to cross me off the list. I even remember saying, "I wouldn't have even been able to go, so she really should have invited me."

My sister called me on my attitude, asking, "When was the last time you picked up the phone and talked to her?" When I admitted it had been a while, she replied, "What do you expect, Sam? You gotta make an effort. As Nanny always said, 'The grass is greener where you water it.'"

Social Relationships Cannot Be an Afterthought

Happiness comes from "with," not within. Sadly those closest to us are often the first to pay the price when pebbles rain down. Busy schedules, bad moods, fatigue, and mobile devices all contribute to our neglecting the people we care about. What makes matters worse is the widespread belief that mental health is exclusively an individual's responsibility. The motivational-thought industry constantly tells us that whatever we want or need, it's on us to get. Popular slogans that promote this misleading mindset include:

- "Happiness is an inside job."
- "Everything you need is already within you."
- "You have to find yourself first. Everything else will follow."
- "If you want to be strong, learn how to fight alone."
- "You can't go wrong by making yourself a priority."

Instead of empowering us, these mantras weigh us down. As a psychiatrist, one of the greatest challenges I face is trying to persuade patients to focus less on themselves and to cultivate connections with others. Almost every patient wants to be happier, like Suzanne, who literally walked in wearing a T-shirt that insisted TRUE HAPPINESS COMES FROM WITHIN, NOT FROM OTHERS.

On the heels of a breakup with a long-term boyfriend, Suzanne decided she needed to spend more time on herself. Coming to see me was only part of that self-improvement strategy. She also began meditating. She bought a treadmill. She cleaned out her closets and gave away anything that didn't "spark joy." Every two weeks she would detox with a juice cleanse. When I asked her how often she saw her friends, she responded, "Not so much these days." She was making herself a priority.

The belief that self-centered endeavors are the only way to access well-being is misleading and often harmful. Social relationships cannot be an afterthought. A sense of belonging is not incidental to our positive mental health. It's central.

I told Suzanne about a study involving people who were asked, "What can you do so that you will be more satisfied in the future?" Their answers included: stop smoking, get a better job, make more money, spend more time with friends and family, and help people.

A year later, the researchers checked back in with the study's participants. Responders who described pursuits that involved other people experienced more positive changes and were more satisfied with their lives than those who described nonsocial strategies. Social engagement trumped personal gain.

The study's authors concluded: "Beyond new insights into how people can potentially increase their happiness, our research underscores the crucial role that other people play in our lives. Social relationships and affiliations have powerful effects on health; loneliness and social isolation have even been associated with increased mortality. Our analyses further strengthen the notion that investments in social relationships are an important leverage point for achieving a healthy, long, and *happy* life."

Similarly, exercise that involves more than one person yields the greatest boost to well-being. People who engage in team sports report having the fewest days of poor mental health each month. Getting physical with others might even help you live longer, too. According to a study of Danish men and women, tennis players (associated with an extra 9.7 years), badminton players (6.2 years), and soccer enthusiasts (5 years) enjoy longer lifespans than people who engage in solitary activities such as jogging (3.2 years), swimming (3.7 years), or cycling (3.7 years).

Raising your heart rate is good. Raising it with others is even better. Plus, scheduling a workout or walk with a friend reduces the chances of your skipping it. Behavioral scientists call this a commitment device. I call it not wanting to be "a flake." I have a standing date to take a walk with a friend on Friday mornings, and without fail there always comes a moment when the thought crosses my mind to text her that I need to cancel. There are a million reasons I manage to find to skip our meeting: it's too cold, it's too hot, I'm tired, I have a dozen emails to return, I just don't feel like it. But not wanting to be a flake who disappoints a friend gets me out the door every time.

By its very nature the busyness of everyday life draws us into ourselves and away from others. The daily grind invites self-immersion. Hassles promote self-focus. This tends to be corrosive on its own and can be exacerbated by the conviction that well-being is exclusively an internal and individual process. Retreating into oneself and self-focus promote narcissistic behavior, which is the opposite of love and connection. "Other people matter." Renowned positive psychologist Christopher Peterson's famous line captures what decades of research have discovered: deep and meaningful close relationships are at the heart of well-being. Lacking positive relationships is as big a risk factor for early death as smoking cigarettes, alcoholism, obesity, and air pollution.

As a society we value independence and prize self-sufficiency. We celebrate individual achievement and champion personal happiness and success. Positive attributes like motivation, self-control, confidence, and curiosity are assumed to be internally generated. When we speak of resilient people, intrepid individuals who overcame major challenges come to mind—John McCain, Malala Yousafzai, Oprah Winfrey, and Bethany Hamilton. We prize their inner strength and personal fortitude. And when we consider everyday wins or setbacks, we chalk those up to an individual's personal qualities as well.

Mary did so well on her exam because of her motivation.
Jack can't stick to his diet because he lacks willpower.
Jackie gave a great speech because of her confidence.

A singular focus on the individual oversimplifies the more nuanced story of what might efficiently and effectively enable someone to reach his or her potential.

It is easy to emphasize the power of individual achievement, but the reality is, every great success story is the product of a huge network of collaboration. Parents, mentors, teachers, coaches, and a web of supporters all contribute to a person's particular greatness. Even in fields that appear to be dominated by singular greatness, the critical roles that others play are undeniable.

In the game of golf, for example, we know the names of the greats—Jack Nicklaus, Tiger Woods, Arnold Palmer—but only a superfan knows the names of their caddies. And yet each caddie is instrumental to the success of the champion golfer. Good caddies function as philosophers, friends, psychologists, and coaches. They are partners who help a player stay focused and build confidence, especially after setbacks like missed putts. Caddies play a key role by helping players to relax under pressure and not overthink their shots. High fives, bear hugs, and a well-timed joke can help a player perform at his or her peak. As professional golfer Johnny Miller observed, "I don't think anywhere is there a symbiotic relationship between caddie and player like there is in golf."

To forge connections is a key to attaining success in virtually every profession. In the art world it has long been held that Johannes Vermeer was a lone genius, but an exhibition at the National Gallery of Art in Washington, DC, dispelled the myth. Vermeer was deeply engaged in an artistic milieu, and his work was fueled and inspired by the exchange of ideas with his peers. Supportive relationships promote personal growth and serve as tailwinds. Psychologists at the University of Michigan found that the more supported people feel, the greater their confidence to take on a challenge.

My patient Gina was an entrepreneur who built a wildly popular organic brand. Her former high school wanted to showcase her success and invited her to speak to a graduating class during its commencement. Gina felt honored, yet her first reaction was to reply, "No way." Writing a graduation worthy speech would take a lot of time and effort, and Gina didn't feel up to it. She went back to the committee and offered, "Why don't I just attend the ceremony and the cocktail party the evening before? That way I can meet students and answer any questions they might have."

Gina's husband recognized that she was dismissing an opportunity to reach many more students with her message. He convinced her to meet the challenge, reminding her about the great speech she had given at a conference the year before. He offered to read drafts and to listen to her practice. She finally agreed. Gina's speech was a hit—full of inspiration and humor. She felt thrilled that her husband had encouraged her to push herself. The best relationships serve as sources of strength during tough times. And in the absence of adversity, they also encourage us to embrace, pursue, and fully participate in opportunities that enhance well-being and foster growth.

On the surface, people described as "go-getters" are assumed to possess an abundance of internal qualities that help to spur them on. But the confidence to tackle a challenge and pursue a goal is also generated from sources well beyond the individual. Many tributaries pour into the river that ultimately constitutes the self. With this in mind, in addition to admiring others for their achievements, perhaps we should also thank their parents/friends/partners for offering love and support. Behind every resilient person there is usually someone who believes in them and has his or her back.

Hills Less Steep to Climb

When you feel socially connected, little things are less likely to get under your skin. Studies show that a hill feels less steep to climb when a hiker is accompanied by a friend. Public speaking seems less stressful after a hug from a loved one. A study of paramedics found that even after a high-stress day—such as one in which violence was encountered, a patient was lost, or a call was received that involved a child—those who reported high levels of perceived social support slept much better than those who reported low levels. Feeling loved and cared for kept their sleep quality relatively stable despite the challenges of their hectic lives.

Another study of married couples proved that the simple gesture of holding a spouse's hand can buffer physical discomfort. Hand-holding reduced not only the subjective perception of pain (in this experiment, how much an electric shock hurt) but also the brain and the body's physical response to pain. Holding the hand of a stranger was better than to hold no one's, but the quality of the relationship mattered: the happiest couples experienced the greatest benefits. Knowing you are loved literally lessens pain, lightens the load, and propels you forward up those steep hills.

Chapter 12

GAIN SOME DISTANCE

"Who is the real me?"

In our first session Kelly described her daily life as "crisis surfing" and said she had entered therapy to "find herself." She'd already begun the process by reading a lot of self-help books and prioritizing self-care. Through self-reflection, she hoped she might finally be able to figure out who she was and what she wanted. And then success, happiness, and fulfillment would follow.

When you max out your mental capacity, self-focus can be a form of self-preservation. It is a reflexive response when life feels out of control and everyone seems to want a piece of you. There are also powerful cultural and commercial messages that support this approach, such as those from wealthy, gorgeous celebrities.

When you're pelted by the pebbles of everyday life, to focus on the self is a tempting option, but it's not an optimal strategy for feeling better. Self-focus might make you feel less vulnerable, but it can also seal you off from vitality. Spiraling inward, going over what is bothering you again and again with a fine-tooth comb, scrutinizing every little detail of what has happened, or might happen, can play a role in the onset and maintenance of depression.

Instead of self-immersing, put some space between you and whatever is bothering you. Self-distancing helps to take the sting out of everyday upsetting events. Failures and setbacks seem less personal. Self-distancing can even help you to better manage anxiety about future worries, such as failing an exam or getting sick.

Establishing mental separation from an emotional situation gets the ego out of the way, provides perspective, and makes it easier to move forward. It also can decrease rumination. The word *rumination* comes from the Latin term *ruminari* and means to chew cud—partially digested food that is regurgitated from the stomach for another round of chewing. In fact, the first stomach compartment of ruminants (cattle, deer, giraffes) is known as the *rumen*. When we ruminate, we are mentally chewing partially digested thoughts.

Participants in one study were asked to recall an intense negative experience and then to adopt either a self-immersed or self-distanced perspective. The self-immersed group were given one set of instructions: *"Go back to the time and place of the experience you just recalled and see the scene in your mind's eye. Now see the experiencing unfold through your own eyes as if it were happening to you all over again. Replay the event as it unfolds in your imagination through your own eyes."*

The self-distanced group were given a separate set of instructions: *"Go back to the time and place of the experience you just recalled and see the scene in your mind's eye. Now take a few steps back. Move away from the situation to a point where you can now watch the event unfold from a distance and see yourself in the event. As you do this, focus on what has now become the distant you. Now watch the experience unfold as if it were happening to the distant you all over again. Replay the event as it unfolds in your imagination as you observe your distant self."*

Both groups were then asked to analyze their feelings about the experience. The participants who were cued to analyze their feelings from a distance focused less on *recounting* the emotionally charged features of the negative experience and more on *reconstruing* it in a productive way that gave them insight and closure. With distance, they didn't just rehash what happened to them. They processed it.

One self-distancer wrote, "I was able to see the argument more clearly. . . . I initially empathized better with myself but then I began to understand how my friend felt. Those feelings may have been irrational, but I could understand his motivation."

This response reflects a desire to move on as the self-distancer replaces the pain surrounding the argument with understanding, which allows her to feel more empathy. Compare this with the raw and agonizing response of a participant who was asked to analyze her feelings about a negative experience from a self-immersed perspective: "Adrenaline infused. Pissed off. Betrayed. Angry. Victimized. Hurt. Shamed. Stepped-on. Shitted on. Humiliated. Abandoned. Unappreciated. Pushed. Boundaries trampled upon."

Immersing yourself in your feelings does not necessarily facilitate insight or understanding. It can actually get you stuck. When you step outside of yourself, it is easier to understand opposing viewpoints and appreciate subtlety. Issues are less black and white. A bit of distance can help you make better decisions and manage negative feelings more adroitly.

Immersing yourself in your inner thoughts narrows the lens through which you view the world—think of a horse putting on blinders. That perspective may be ideal for barreling ahead but terrible for viewing the entire landscape. Frequently discussing and speculating about yourself and rehashing your problems can increase distress and undermine resilience.

Let's say you're having a conflict with a coworker. When self-immersed, you're more likely to jump to a superficial explanation for the disagreement. You might simply dismiss your colleague as an "idiot." If you're self-distancing, you'll gain the perspective of the entire landscape and be able to identify where your views actually diverged. You might still conclude your coworker is annoying as hell, but it won't feel as personal.

When you are self-immersed, it is more difficult to be coolheaded and put negative issues aside. Similar to having a sportscaster in your head providing play-by-play commentary, every annoyance and irritation tends to grab your attention. A literal chain of frustrating events—"My dog threw up on the carpet this morning, then my five-year-old stepped in it wearing his brand-new dress shoes we had bought for my cousin's upcoming wedding that I don't even want to go to, and then he walked on our brand-new carpet, and then . . ."—adds up and amplifies the accompanying negative emotions. Brooding makes it hard to concentrate on anything else. Your mood takes a nosedive. Anger bubbles to the surface. The experience of fatigue or any bodily aches and pains that might otherwise have gone unnoticed is now a constant.

Taking an outsider's point of view can provide insight and reduce negative feelings about setbacks and conflicts from the past. Self-distancing also has application in everyday life. On those days when everything seems to be going wrong and you are convinced that the world is conspiring against you, remember . . . it's probably not. When you take everything personally—the endless red traffic light, the five people who get into the elevator with you and press *every* floor below yours—emotional stamina is depleted.

Stepping outside of yourself can help reduce irritation and anger in the moment. A patient of mine is a huge Elvis Costello fan. When he'd feel irritation mounting, he'd start to sing to himself the lyrics from the song "Red Shoes," "I used to be disgusted, but now I try to be amused."

In a study, "Flies on the Wall Are Less Aggressive," researchers intentionally provoked participants by interrupting and berating them for not following directions. Those who were instructed to view the takedown from an emotional distance exhibited less anger than those who immersed themselves in their emotions. "The worst thing to do in an anger-inducing situation is what people normally do: try to focus on their hurt and angry feelings to understand them. . . . It keeps the aggressive thoughts and feelings active in your mind, which makes it more likely that you'll act aggressively," explained the coauthor of the study.

Self-distancing isn't just good for you, it's good for your relationships. Going beyond our own issues helps us get along better with others. There are so many opportunities to use self-distancing—to stop us from ruminating about something that happened in the past, to provide perspective in the moment, and to quell worries about the future.

Time Travel

Another way to disrupt rumination is to time travel. Imagining what your future self might think about a current stressor has been shown to reduce the emotional toll of the present. For example, as upsetting as an interaction with that annoying coworker might be today, fast-forwarding from the current situation to a year in the future might help you take it less personally and see it as less permanent. Recognizing the transitory nature of a hassle can reduce the distress you feel about it.

Projecting into the future can also help people overcome relationship conflicts. Participants in a study were asked to think about a recent conflict with a partner or close friend. The experimental group was asked to consider how they would feel about the conflict a year later. The control group was asked to describe how they were feeling currently. The researchers found that those who considered the conflict from a future perspective showed more forgiveness and greater insight. They also reported feeling more positively about their relationship. This strategy worked wonders with my patient Chiara, who had trouble letting go of grievances. She had an uncanny ability to remember the specifics of any argument, slight, or disappointment, and cataloged them like a librarian trained in the Dewey decimal classification system, so they were ready to be recalled at a moment's notice. I could see her visibly tense up when she spoke about them, ensnared in a web of overwhelming emotion.

Adopting a future perspective enabled her to view these moments without all the messiness.

"It's like I had the windows cleaned," she explained. "No more grime in the corners, and I can see out perfectly."

Talk to Yourself

Changing how you talk to yourself is another way to stop yourself from tumbling down the rabbit hole of relentless self-focus. Before you roll your eyes, let me make it clear that I am not suggesting talking to yourself out loud. Nor do I endorse referring to yourself in conversation in the third person, such as in, "Samantha prefers her pasta al dente." But there is evidence that using your name and third-person pronouns (e.g., he, she, it, itself, they, them) to refer to yourself in *silent* introspection during a stressful moment can enhance your ability to control your thoughts, feelings, and behavior.

For example, participants in a study were told they had to deliver an important speech about why they were ideally qualified to land their dream job. They had just five minutes to prepare it. As you can imagine, it made for a stressful situation. When the five minutes were up, half were asked to reflect on their anxiety using non-first-person pronouns or their name (i.e., "Is Samantha going to give a good speech?"). The other half were asked to reflect using first-person pronouns ("Am I going to give a good speech?"). Next, each participant was asked to deliver his or her speech in front of a panel of evaluators. The speeches were also recorded and shown to judges who knew nothing about the conditions of the experiment or the instructions the participants had received. The participants who had been asked to reflect on their anxiety using non-first-person pronouns were judged to be more confident, less nervous, and better overall than those in the first-person group. Moreover, participants in the non-first-person group said they felt less embarrassed and reported ruminating less about their performance afterward.

A patient with social anxiety realized that whenever she spoke to herself using "I," she became hypercritical and riddled with doubt. Like the chyron at the bottom of television news channels, her inner voice relentlessly informed her, *I'm not up to this. What's wrong with me? I'm so awkward. I should just go home.*

I encouraged her to try speaking to herself in the third person to see if that perspective shift led to a healthier inner dialogue. Instead of opting for "Chiara," though, she chose a new name for addressing herself: Wonder Woman. She told me that when thrown into anxiety-provoking social and professional situations—meeting friends at a bar, the morning of a conference, before an interview—she would look in a mirror and say, "Hey, Wonder Woman, you got this."

This story delighted me. She not only shut down her self-loathing voice, she boosted her confidence.

What Would You Tell a Friend?

Many of us are good at offering sage advice to others, but when it comes to our own personal dilemmas, we're useless. This disconnect actually has a name—"Solomon's Paradox," after the king of ancient Israel, Solomon, who was known for his wisdom. His subjects traveled far and wide to seek his advice, but in his personal life, the king was a hot mess. Granted, it's hard to juggle hundreds of wives, especially when you add in concubines, too. Still, King Solomon's biggest failure was as a father. He had only one son, who grew up to be an incompetent ruler, which led to the kingdom's downfall.

When we are locked in our thoughts, it's harder to see the long view. We are more likely to focus on the minutiae and the immediate consequences and less likely to consider the broader ramifications of our actions or to be aware of our biases. Self-interest can involve more than putting up blinders—it can be blinding.

Considering how you would advise a friend in a similar situation can help you make a better decision for yourself. When you make a problem less about you, you ultimately see more. Research shows that "decentering"—shifting the focus away from yourself and toward someone else—promotes more effective conflict resolution and more effective reasoning about personal issues. Decentering is also linked with cultivating greater humility and an awareness of one's own shortcomings, and with feeling greater appreciation for another person's point of view.

I often encourage patients to use self-distancing when talking or thinking about something that upset them in the past and to deploy these strategies to deal with everyday challenges. I also urge them to be mindful about how they speak about their problems with their friends.

A patient who had a difficult relationship with her mother-in-law would spend hours rehashing the latest incident or insult with her best friend and speculating about what her "monster-in-law" might do next. It was their go-to conversation, or as my patient called it, "a bitch fest." Afterward, she would often feel even angrier.

Venting can feel great in the moment, but doing so repeatedly without any resolution or forward progress can make you feel worse. Excessive complaining and rehashing personal problems with someone else is known as co-rumination and can amplify stress, especially in those who are already feeling down. If your best friend calls you to talk about something that is bothering her, it is best to avoid questions that encourage her to revisit every detail. "Start from the beginning. Tell me everything!" will only lead to a play-by-play of what took place and what she was feeling. Consider instead posing a question that might help your friend gain some distance from the situation. I often ask my patients, "If someone else were in this situation, what advice would you give them?" Rather than dwelling on the details, help others generate a plan of action.

It is natural for parents to want to know what's going on in an adolescent's head. But a parent who fixates on what's going wrong might do a child more harm than good. In a study of four hundred students—fifth, eighth, and eleventh graders—researchers found that adolescents who co-ruminated with their mothers were more likely to co-ruminate with their friends and to become anxious and depressed as a result. Circular conversations about negative feelings, without resolution, fuel worry and feelings of sadness and can lead to indecision. Undergraduates who co-ruminated with their parents were more likely to suffer from anxiety. It is likely that co-rumination fosters the unhealthy tendency to ruminate on one's own.

My patient Helena told me that when she was a child, her mother would read the newspaper every morning at breakfast, and whenever she came upon a criminal or tragic story, she would gasp and read the full story aloud, dramatizing the goriest details.

"Listen to this," she would say. "Oh, it's just awful. Can you imagine? An entire family wiped out in a fire? How horrendous!"

When Helena's father entered the kitchen, there was a repeat performance. And that wasn't the end of it. If her mother bumped into a friend at the bus stop, she would bring the story up again: "Did you hear about that awful fire?"

Rumination was this woman's way of bonding with people, including her daughter. She recalled how the two of them would sit together in the car on the way to school rehashing complaints, mulling over grievances. Commiserating with her mom created a fleeting feeling of closeness but didn't teach Helena how to manage discomfort or face a challenge.

As an adult, Helena adopted a similar appetite for conversations that dwelled on misery. Eventually, she realized how co-ruminating left everyone involved stuck in an emotional sinkhole without a way out. These discussions created pseudo-bonding, eliciting emotion that was superficial and disconnected to people's actual lives. Helena eventually realized that whenever a conversation began to feel as if she had had it before, it was a sign for her to change the subject.

It's remarkable how easily one person's conversational style can rub off on another. The results of one study indicated that a short chat with a ruminating stranger can induce rumination, even in someone who doesn't typically ruminate. For better or for worse, we mirror the behavior of the people around us. And when it comes to ruminating, it is definitely for worse. If the person sitting next to you on a long flight breaks the ice by asking, "Do you ever wish you had made different choices in your life?" steer the conversation in another direction or put on your headphones.

Something Bigger Than You

If your teenager is upset about something, asking her to recount every little detail to you and perhaps later to your partner—"Tell your stepmom exactly what happened today at school"—could make her feel even worse. Turning every problem into an ongoing centerpiece of conversation can ultimately undermine resilience. You might be sending the inadvertent message that the issue is worse or more serious than it is, or that you believe your child can't handle the situation. If that child had an issue with Johnny today, don't let the first question you ask after school tomorrow be "Was Johnny mean to you again today?" If you don't let it go, your child won't either. When listening to your adolescents, stay calm, express empathy, and encourage them to consider the situation from alternate perspectives. These are more effective methods for working out problems than mining for pain and dwelling on anger.

Instead of making a teenager's problems the go-to conversation, consider that kids who know a lot about their family history—where their grandmother went to school, where their great-uncle grew up, how and when their parents met, and so on—have been shown to be more confident and resilient overall than those who don't. Researchers theorize that helping children develop a strong "intergenerational self" reminds them that they belong to something bigger than themselves. They are not the central character in the family story but part of a chapter that is still being written. It is the opposite of "all about me" and as such has a valuable decentering effect.

Knowing that the family made it through challenging times in the past—that its story continues to unfold—makes the setbacks adolescents encounter feel less daunting. Rudyard Kipling captures this perspective perfectly in his poem "If." When you know the arc of your family narrative, you can meet with "Triumph and Disaster / And treat those two imposters just the same."

When you are wrapped up in yourself, it is more difficult to spot or capitalize on what exists beyond you. It is harder to accept another person's perspective, and uplifts are scarcer. Apprehension, pressure, tension, worry, and the dread of what's next draw us into ourselves. There are many reasons why people engage in resilience-sapping behavior. It is reflexive. It is what others are doing. It is momentarily self-soothing. But there are more healthy responses to distress. We can get closer to vitality by gaining some distance from ourselves.

Chapter 13

TAILWINDS

Human Beings are made up of each other. We are deeply social creatures.

Our bodies constrain us, but our social interactions make us who we are.

TANYA LUHRMANN

When I ask patients to tell me about their day, they inevitably speak about the people in their lives. They open up about their exhausting but wonderful kids, their narcissistic colleagues, their hard-to-read boyfriends, their hot-and-cold mothers-in-law, their hard-to-please teachers, their always-there-for-them friends, their adoring-but-needy grandparents, their life-saving babysitters, and their entitled bosses.

Positive, uplifting interactions are often considered the best part of a day. We are grateful to the person who made us laugh. We cherish a sweet text from a loved one or a compliment from a coworker. Even a random interaction with a stranger—a barista, a salesperson, someone in line at the post office—can make the list of a day's highlights.

Likewise, negative interactions are a significant source of everyday distress. It is no surprise that an argument is one of the most upsetting of daily aggravations. A harsh exchange in the morning can cast a shadow over the rest of the day. An unpleasant conversation with a colleague, or even a stranger, can leave one feeling ill at ease. Our interactions with others and the emotions they uncover reveal a lot about who we are. Transference is a concept first introduced by Freud in his book *Studies on Hysteria*. Transference occurs when feelings about one person are redirected toward another. Noting the deep and intense feelings that patients developed toward him in therapy, Freud postulated that the patients were unconsciously projecting their feelings about important people in their lives onto him. I'm not a Freudian analyst, but I am deeply interested in my patients' interactions with the important people in their lives as well as the seemingly minor ones.

One patient in her late forties, Beatrice, spoke at length to me about "a jerk" who didn't hold the elevator for her on the way to my office. This small act unleashed a stockpile of anger and triggered my patient's anxiety about feeling invisible. She took the slight personally, convinced the man would have held the elevator if she'd been a beautiful twenty-year-old.

I asked her what I often ask patients to challenge their certitude: Is there an alternative explanation? It is certainly possible the man would've only held the door if she'd been a supermodel, but it was also possible he was late for an appointment and in a rush . . . just like her.

We often look for evidence that reinforces our beliefs and confirms our fears. Any ambiguous situation can serve to strengthen our convictions. When the elevator door dis happened, Zara was already dreading an upcoming birthday after reading an article that claimed a woman is "most beautiful at the age of 36." "From then on," she half joked, "it's all downhill."

Zara was concerned about growing older, but vanity wasn't the issue. As we talked more, it became clear that she was afraid her husband would leave her for a younger woman. Zara had witnessed this dynamic in her parents' marriage when her father left her mother at age forty-five for his young secretary. Zara rolled her eyes at how clichéd the situation was, but it haunted her. While she sympathized with her mom, she also blamed her for not trying harder to hold her father's interest.

Beatrice acknowledged that during the past few months she had become increasingly cold and distant with her own husband. Her brief encounter with an impatient man unearthed her conviction that it was inevitable she would repeat her parents' history and it would all be her fault. Surfacing this fear enabled her to address it directly.

I spend a lot of time talking with patients about everyday encounters—the good, the bad, the ones they wished they had, the ones they intentionally avoided, and even the ones they may not have even realized they missed. It is not that patients don't value their connections. They do, but the daily grind gets in the way. They forget to make the effort to connect.

Positive relationships support individuals not only in their ability to cope with stress and adversity but also in their efforts to learn, grow, and explore. Simply put, they are a reservoir of vitality. Given the benefits of meaningful relationships in one's life, it's worth taking a closer look at what actually constitutes a good one. In an influential study, "The Need to Belong," social psychologists explored the nature of social connections and the underpinnings of lasting and meaningful bonds with others. Their findings showed that a critical aspect of any healthy and fulfilling relationship is to have frequent positive interactions. This means, for example, that being in the same room is not enough if everyone in the family keeps his nose in a screen. In the context of friendship, Liking a friend's Instagram post, retweeting her tweet, and sending a group email wishing everyone on your contact list a Happy New Year doesn't count: they're low-effort connections that offer a low emotional payoff.

You can have many daily interactions with people and still feel lonely. A bond in name only lacks the mental and physical health benefits of a high-quality connection. When interactions are ridden with conflict and contempt, or if they feel empty and devoid of care and concern, then it's a relationship, but not a satisfying one. One patient tells me she sometimes has to text her boyfriend to get his attention even when he's sitting right next to her. In fact, two in five Americans sometimes or always feel their relationships are not meaningful.

Having social connections does not necessarily mean you *feel* connected. Feeling cared for, appreciated, and understood are necessary for a sense of belonging. It is often the subtle behaviors rooted in the daily fabric of relationships that communicate a genuine sense of love.

Married couples who lived apart (such as so-called "commuter couples") revealed it was "the little things" they missed most about living together—those seemingly trivial moments like sipping coffee together in the morning or watching *Jeopardy!* together in the evening. Frequent long-distance phone conversations helped but didn't make up for the lack of day-to-day interactions. Sharing information and discussing practical issues didn't produce the same pleasant feelings associated with enjoying each other's company.

One might assume the main reason commuter couples are less happy than those who live together is that they have less sex. The evidence suggests otherwise. Most of the commuter couples in the study typically had sex only during weekends, even during the times when they lived together (which is consistent with surveys of cohabiting married couples), so being apart during the week was unlikely to impact their sex life.

Beyond sex there is something about being together both emotionally and physically that is necessary to satisfy the need to belong. It is in the warmth, the eye contact, the hand on the shoulder, the warm hug, the peck on the cheek. Indeed, such "insubstantial" interactions turn out to be quite substantial. According to the UK Campaign to End Loneliness, more than half of lonely people simply miss having someone to laugh with. It is the simple everyday activities, such as sharing a meal (35 percent), getting a hug (46 percent), holding hands (30 percent), and taking country walks (32 percent) that people long for.

Simple everyday positive interactions are the lifeblood of companionship and connection. There is some evidence that these interactions are more valuable than overt gestures of support or concern. I had a well-intentioned friend who would always make assumptions about my well-being. Whenever we saw each other, she would offer support with just a tad too much concern.

"Is everything okay? Be honest. Are you *really* okay? You look tired. I'm going to get you a kombucha ginger shot to help you combat that cold you're probably coming down with," she said with furrowed brow.

I actually felt fine until she commented on my bedraggled appearance. Then I couldn't think of anything else: *Do I look that bad? Is it time to pound kombucha? Maybe I am sick . . . ?*

Support can do more harm than good when it's out of sync with another person's needs. Before you voice concern or offer help, take into account how it might impact the person on the receiving end. Sharing unsolicited advice can inadvertently make them feel inadequate or incompetent. Think of the parent who swoops in to help with the child's math homework before the child even asks a question. Being overinvolved can undermine someone's confidence. Be responsive to the person in a way that helps them feel understood and validated, and that shows you care—with or without kombucha.

Invisible Support

Unlike heavy-handed gestures, the most effective kind of support is often invisible to the recipient. For example, you might shield your partner from having to worry about a broken dishwasher by managing the repair process yourself. You might tidy up so that your partner comes home to a calm and peaceful apartment. If a partner has a long day and needs to use the car first thing in the morning, you can make sure that it has a full tank of gas. If your partner's boss sends a weekend email asking for a meeting first thing Monday morning, framing that as a positive can shape whether your partner views the message as a threat or an opportunity. If a school friend is worried about an upcoming math test, you could ask the teacher to hold a study session for the entire class.

Using a daily diary, law students preparing to take the New York State Bar Examination were asked to record their experiences of receiving emotional support from their partners in the weeks leading up to the exam. Their partners were also asked to record acts of providing support to the student. The students who received invisible support—support that was not obvious to them—experienced less anxiety and stress than those who received more overt support. It seems that the most effective kind of support takes place without drawing attention to the recipient's specific needs. It is possible that providing invisible support is the essence of being a good parent, mentor, friend, partner, and doctor.

Micromoments of thoughtfulness are the secret sauce of robust relationships and contribute to feelings of connection and greater well-being. The reason people with strong social ties are healthier and feel happier and more vital is not because they are more likely to ask for and receive support whenever they feel the least bit of tension, but because their loved ones are more likely to enhance their lives in countless subtle ways.

Communicate Your Affection

Flowers and chocolates are nice, but it's the implicit communication of affection that makes a person feel loved. Responsiveness is essential for expressing that affection; it's a bedrock of intimacy. Responsiveness entails consistently showing another person that they are genuinely understood, valued, and cared about every single day. Tender behavior leads to high-quality interactions that increase love and passion in both the short and long term.

During the beginning of a relationship, couples cannot keep their hands off of each other. Romance and sexual desire take center stage; then, over time that first blush of passion fades. The assumption is that desire diminishes as familiarity and intimacy set in. This is known as the "intimacy-desire paradox." The paradox is predicated on the belief that desire is driven by uncertainty and novelty, so the more emotionally connected people are, the less desire they feel for each other.

But a study conducted in Israel has called the intimacy-desire paradox into question. The research shows sexual desire actually *thrives* with rising intimacy. "Being responsive is one of the best ways to instill this elusive sensation over time," wrote researcher Gurit Birnbaum, "[It's] better than any pyrotechnic sex."

Behaving thoughtfully and making an effort with each other fuels sexual desire, especially for women, although men report a boost, too. Bringing home her favorite ice cream, taking turns choosing movies, deciding to read the same book, and sending him a flirty text are among the many ways you can be responsive. Of course, not all intimacy is created equal. Brushing your teeth together, leaving the bathroom door open, and clipping your toenails in front of your partner might not fan the fires of desire. As the saying goes, if you act like you did at the beginning of the relationship, there wouldn't be an end.

When there are kids to feed, bills to pay, and laundry to do, being nice doesn't always come easily. Taryn realized she was taking out her frustrations from her long workdays as an emergency room doctor on her partner and kids when she got home. "I storm into the house, pointing fingers and barking orders: 'Julian, get in the shower NOW.' 'Ethan, why aren't you studying for your history test?' 'Why hasn't anyone walked the dogs? Do I have to do everything around here?'"

Because of the way Taryn issued such commands, her husband and kids started to refer to her as "Stormin' Norman," after US General Norman Schwarzkopf Jr. On her arrival home, she and her husband would begin bickering, the kids would start fighting, and even the dogs would get riled up. In therapy, Taryn recognized that her husband wasn't thoughtless, as she accused him of being, and that her kids weren't the ingrates she complained about. The actual issue was that pent-up frustrations from the day were infecting her evenings at home.

We discussed creating a "decontamination protocol" to help her "wash off" the negative residue of the day. I told Taryn about a high school English teacher who would sit in her car for fifteen minutes and listen to heavy metal music to decompress. I knew this tactic wasn't going to work for her, but it sparked a few ideas, such as instead of scrolling through work emails on the train on the way home, Taryn decided to read a novel; she got off the subway a stop earlier so she had more time to walk off the negative energy; and she channeled a scene from the movie *Outbreak*: the moment Taryn got home, she changed out of her work clothes and into soft sweatpants.

I also encouraged Taryn to devise ways to manage stress more effectively at work so it wouldn't take such a toll at the end of the day. She had always considered taking breaks a waste of time, but discovered they actually made her more productive. Her kids came up with a third idea: they took her phone and changed her screen saver to an image of General Schwarzkopf. Though he was a fine commander, they explained, they didn't like it when their mom morphed into him. The image on Taryn's phone made her smile. Laughing at yourself is always good medicine.

How Happy Couples Argue

When I was younger I assumed that people in happy relationships rarely fought. During my early years of clinical training I asked an elderly patient, Margaret, who had been happily married for over fifty years, how she and her husband avoided conflict.

"Are you joking?" she asked, howling with laughter "We argue all the time!"

It turns out that happy couples have disagreements, just like everyone else—what's different is *how* they fight.

"Even when we agree to disagree, I always feel like he tries to take my perspective and see the situation from my side," Margaret explained.

Again, responsiveness is key. Sensing that your romantic partner is trying to understand where you're coming from and respects your point of view helps to protect a relationship from the potentially harmful effects of conflict and can even strengthen your bond.

When happy couples argue they stay on topic, steering clear of generalizations like, "Here you go again." Insulting extended family, especially mothers-in-law, is a definite no-no. "Why" questions—as in "Why are you so upset?"—are also to be avoided, because they invite blame and defensiveness.

Asking a "what" question—as in "What's on your mind?"—is less confrontational. Happy couples also know how to pick and choose. Instead of dwelling on issues that aren't easily changed, like one's partner's snoring or an in-law's sarcasm, they focus on problems that are more easily resolved. Margaret told me she always selected her battles wisely. "You can argue about every little thing until the cows come home, but what's the point? If it's not solvable, we don't bother."

It is important to show your partner you're interested in what he or she is saying when you argue. It is even more important to demonstrate interest when they share good news, according to Shelly Gable of UC Santa Barbara. For instance, imagine Maria comes home from her job as an associate at a law firm and excitedly tells her husband she has been assigned to be the lead lawyer on a big case. He could respond in a number of ways, including:

1. With wholehearted enthusiasm: "Wow. That's awesome! What's the case about? Your hard work is really paying off. I am so excited for you. Tell me more."
2. Like a distracted robot: "Cool," he mutters while checking email.
3. Rain on the parade: "It sounds like it's going to be a lot of work. Do you think no one else wanted the case?"
4. Hijack the conversation: "Guess what happened to me today!"

Wholehearted enthusiasm is known as "active constructive responding," or ACR, and it's the only style associated with higher relationship quality and greater personal well-being. Expressing interest and support brings couples closer and enhances connection, while the other three styles are all negatively associated with relationship quality.

Active constructive responding benefits all types of relationships, including with friends, students, coworkers, and children. Listen carefully the next time someone you care about reports something good that has happened. It may be as simple as their showing you an article in a newspaper that interests them or telling you about a book they just finished. Give this person your full attention. Look up from your phone. Ask questions. Relive the moment with them. If you're at a loss for words, just remember these three: "Tell Me More."

Of all the research I learned about while studying positive psychology, Gable's work had the most immediate impact on my everyday life. I cringed when I first encountered it. Although I didn't hijack conversations to make them all about me, I was definitely guilty at times of being a distracted robot, barely grunting acknowledgment when my husband tried to tell me something. (*Can't he see I'm busy?*) Without intending to, I sometimes rained on my kids' parade. Once, my daughter excitedly told me she wanted to join the basketball team. "Okay," I responded. "I just hope it doesn't interfere with your schoolwork. When are the games? Are you sure it's a good idea?"

I wasn't just raining; I was thunderstorming. After reading Gable's research, I made the conscious decision to become an active constructive responder. It doesn't always come naturally. On a recent evening my husband mentioned he had seen a "fantastic" exhibition of a well-known artist's late works. "Really," I was about to reply, "I haven't heard the best things about it." Instead I bit my tongue and said, "Tell me more."

Responsiveness is a catalyst for "felt love." It is well established that positive emotions promote health and well-being in individuals. Shared ones are even more powerful. Make the most of them.

Chapter 14

BETTER CONVERSATIONS

> Ah, good conversation—there's nothing like it, is there? The air of ideas is the only air worth breathing.
>
> EDITH WHARTON, *THE AGE OF INNOCENCE*

A few years ago I attempted to engage my children—then, ages nine and ten—in what I believed to be "meaningful" dinner conversation. I had read that the game "Roses and Thorns" was a favorite approach of the Obama family, among many others who had extolled the benefits of family activity.

The idea was to go around the table and each person would recount the best part (the roses) and worst part (the thorns) of his or her day. Maybe Malia and Sasha were more cooperative, but my children didn't take the concept seriously. From the get-go they were skeptical. On day two they rolled their eyes. By day three they were making fun of me: "The best part of my day was recess. The worst part of my day is you asking me these questions!" On the rare occasions when they did manage to generate responses that weren't sarcastic, most of their comments did not lead to deeper conversations. "I got an A minus on my history quiz" (rose). "I forgot my homework at home" (thorn). "I scored a basket" (rose). "I stapled my finger" (thorn). Their contributions were literal and focused on their reactions and their experiences. It was me-centric. Instead of talking about ideas or broader topics, Roses and Thorns gave them the green light for self-immersion.

A few weeks after our failed attempt at Roses and Thorns, I found myself sitting alone in a restaurant, waiting for a friend. A family sat down next to me—a mother and a father with two college-age kids and a younger boy who looked to be in junior high. The tables were what my mother would describe as cheek to jowl. I couldn't resist eavesdropping on their conversation. I just wanted to see how other families did it. It was research, I told myself.

The older son began, "I went to the gym today. I killed it. I can bench press one ninety, no problem." The daughter interrupted, "I didn't sleep well last night. I'm tired and have a headache. I think maybe I'm getting a cold." The smaller one chimed in, "I want the new iPhone!"

The parents tried to shift the conversation to other directions. "Did you hear about what the president did today?" asked the mother. "There was a great article in *The New Yorker* on the new exhibition at the Met," said the dad, trying to pique their curiosity. But the kids had already tuned out and were on their phones, scrolling.

I wondered if this family had played Roses and Thorns over the years. Somehow, like my own children, these kids had gotten the idea that self-interest was the same as general interest. I mentioned these concerns to a friend, Kitty Sherrill, who grew up in the South. She said that when she was a child, the rule at the dinner table was simple: Don't bang on about yourself. It was okay to ask people questions, but it was forbidden to be "Little Miss Guess What Happened to Me Today." Conversation had to be others-focused and ideas-driven.

I was raised to believe there were three topics one should never discuss over dinner: politics, money, and religion. Now those subjects constitute probably the lion's share of table talk. Quality discussions that deepen your understanding of another person or an idea are more gratifying and meaningful. Years ago I attended a dinner at which the artist Brice

Marden was being honored. Instead of giving a long-winded speech, he simply expressed thanks to his brilliant wife: "Thank you, Helen. It's been an amazing conversation all these years."

People who engage frequently in mind-expanding and soul-stretching conversations are happier than those who don't. As one researcher commented, "I would like to experimentally 'prescribe' people a few more substantive conversations."

What separates small talk from substantive conversation? The answer is simple—substantive conversations involve a significant exchange of information. You learn something about the other person or discover information that's beyond the trivial. When we engage in small talk, we walk away unchanged.

Approximately half of Americans (53 percent) report having satisfying in-person social interactions, such as having a nice chat with a friend on a daily basis. Fortunately, having good conversations is a skill that you can build.

Ask Questions

Talking about oneself is most individuals' favorite subject of conversation. One study revealed that people spend two thirds of conversation time discussing their own thoughts, feelings, and beliefs. Many in the experiment were even willing to forgo money so they could chat about themselves. Brain scans reveal that self-disclosure activates areas of the brain associated with pleasure and reward. Put simply, it feels really good to talk about the holy trinity of me, myself, and I. But if you want to connect with someone, ask more questions instead.

People who do ask questions are better liked than those who ask few or none. Question-askers are also more likely to be asked out on a second date and to do well in job interviews. There are many reasons people don't pose questions. Some are unsure of what to ask. Some worry that their questions might be awkward or make others feel uncomfortable. Some are just too focused on themselves. When meeting someone new, focusing on yourself often becomes the default. People try to sell themselves and seize upon any opportunity to redirect the conversation back to their favorite topic. Could we be trying too hard to impress? Researchers found that when applicants went for a job interview "redirecting the topic of conversation to oneself, bragging, boasting, or dominating the conversation, tend to decrease liking." We all know the types who ask about our holiday simply as a means to get us to ask them about theirs.

Asking questions with the intention of learning something, and not just thinking about what you are going to say next, enhances the quality of conversation. That said, asking too many questions or silently listening can be off-putting, too. The best listeners are those who ask questions periodically that promote insight and reflect understanding. The back and forth is essential. We live in a world full of interesting people and ideas—don't let self-interest get in the way of the opportunity to connect with someone or learn something.

To push back against conversational narcissism, Character Lab Education founder Angela Duckworth recommends what she calls "Would you rather?" questions. These are designed to kick-start an engaging discussion. One example is, "Would you rather be extremely lucky or extremely smart?" Or, "Would you rather have people take you seriously or always have people think you are fun?" (In case you were wondering, I'd rather be smart *and* fun.)

Thefamilydinnerproject.org is a source for other thought-provoking questions, such as, "If you were stranded on a desert island, what three books would you bring?" and "What is the one thing you couldn't live without?" Or make a list of your own. If you're at a complete loss, ask a conversation partner what question he or she would most like to be asked.

No Need to Impress

It might be tempting to wow others with a story of an extraordinary experience, like the time you met Brad Pitt, but look for common ground. (And yeah, now that you mention it, I did meet Brad once.)

"Conversations based on ordinary topics have a surprising vitality, while conversations based on extraordinary topics tend to meet an early end. Conversation is something we construct together, and so it thrives on what we have in common. When you depart from this script, don't be surprised if you end up talking to yourself," concluded Harvard social psychologist Gus Cooney.

I had a college-age patient who had just come back from a family trip to Costa Rica over spring break. He went surfing and zip-lining and saw parrots in the wild. When he returned, he assumed his friends would want to hear all about his fabulous vacation and see all his photos. They did . . . to a point. After seeing a few photos their eyes glazed over. After listening to a few stories they lost interest.

"It's not like I'm inviting them over to watch a home movie of my trip; they could at least show some curiosity," he explained. On top of that hurt, he felt left out of conversations they had about the comedy club they visited and the movie they saw while he was traveling.

Like a mosquito bite inflaming the surrounding skin, feeling excluded can quickly swell into a greater feeling of rejection. My patient pulled away from the group, which only made him feel worse. A few weeks later his friends invited him to join them at a bar. To his surprise he had a great time; everything seemed back to normal. They talked, they laughed, and they connected. It was great to learn about what was going on in their lives, and they were curious about what he had been up to. During our next session, this patient told me about his uncle, who was infamous within the family for being a windbag. My patient realized that, upon returning from vacation, "I was acting just like him. And my friends did exactly what I do when my uncle recounts his out-of-this-world experiences—I tune out. He means well and is just trying to share, but I just can't relate."

Because we crave acceptance, camaraderie, and belonging, try to stick to conversations in which everyone feels welcome—not just for your friends' sake but also for your own.

Listen More

The old proverb that children should be seen but not heard was very popular in my childhood home. My sister and I were pretty good at listening, so much so that our parents and their friends would often forget we were there. I remember the alarm on my mother's face when I asked her, "So, what did you mean last night when you said Mrs. X is having an affair with the tennis pro? Does that mean she plays tennis a lot?"

Listening is sometimes the best way to connect and show that you care. As the saying goes, listen to understand, not with the intent to reply. If you ever hear someone disclose something personal and you respond, "I know exactly how you feel," before launching into a monologue, you are in conversational narcissist territory. Sharing is not always caring.

Several years ago I was taking care of a woman in her midfifties with schizophrenia, who believed a tabby cat was living inside her abdomen. She didn't have an issue with the cat—she described Tabitha as friendly and good company, though she wished she wouldn't move around so much. The woman was on the inpatient unit because she had become aggressive when other patients in the residential facility made fun of her and called her "Cat Lady." I had met many patients with bizarre delusions before—one man believed he had been kidnapped by aliens who replaced his red blood with their lime-green blood, while another was convinced he was leaning to the left like the tower of Pisa when he was standing up straight—but this one was particularly unusual. Though she seemed to be in good health and there were no focal findings on her neurological exam, the woman complained of severe headaches, and the doctor from the facility where she lived was concerned that her psychotic symptoms were getting worse. I ordered a CAT scan to rule out a brain tumor. When I told her the plan, her usual lack of expression became animated with emotion. I saw her smile for the first time, and her eyes welled up with tears. "Thank you for listening," she said calmly, and it soon became clear that she thought I had ordered a "cat" scan. The point is: she felt validated. We all want to feel heard and understood.

Be the Dumbest Person in the Room

According to psychologist and intelligence researcher James Flynn, one of the best ways to boost intelligence is to marry someone smarter than you are. He conceives of the brain as a muscle—the more you use it, the stronger it gets. In some of the strongest couples I know, each partner will insist that the other is smarter. What better way to flex your brainpower than to have a significant other who engages you in stimulating conversations and inspires you to stretch your mind and push the boundaries of your imagination?

Communicating about who's picking up the kids and what time the cable installer is coming is necessary, but engaging in soul-stretching conversations enhances vitality.

Chapter 15

LOOK AROUND

Time for a little art appreciation.

Landscape with the Fall of Icarus was painted in the 1560s and perfectly captures the self-absorption often advocated by today's wellness industry.

Attributed to Bruegel the Elder, the painting depicts the mythological Icarus (*lower right*), who fashioned wings made of wax and feathers, which seemed very clever at the time. Filled with pride over his father's creativity, Icarus flew higher and higher until the sun melted the wax and he tumbled into the sea. (The message is not exactly subtle.) Despite this dramatic never-before-seen occurrence, the other characters in the painting appear unfazed. The ploughman plods along, head down, preoccupied with his immediate chores and apparently unaware that a birdman is drowning nearby. The fisherman is so engrossed in his own affairs that he doesn't notice anything is amiss. Another figure stares into the distance, perhaps in self-reflection. A sailboat heads on its way, presumably because the captain was on a schedule and couldn't be bothered to turn the ship around.

The characters are entrenched in their own thoughts, disconnected from one another and their surroundings: there but "not there." Even the horse is oblivious, but at least he has an excuse: he's wearing actual blinders. (W. H. Auden wrote a poem about the collective shrug to the spectacle, "Musée des Beaux Arts," that's worth reading.)

Many of us are often "not there," and this passive, preoccupied existence drains vitality. How often are we so self-absorbed that we are oblivious to the quality of our interactions with others? How often do we miss out on seeing, generating, and participating in vitality-building interactions, experiences, and activities?

The business of living hasn't actually changed that much since the mid-1550s. Yes, we've traded in fishing nets for phones, but ingrained habits, multitasking, cluttered schedules, and fatigue still contribute to a not thereness. We rationalize our self-centeredness in the name of productivity. But, as in the painting, we end up missing a lot and have fewer moments of genuine connection and fewer opportunities to generate positive emotions of any kind. Self-immersion has an emotional impact. It steals joy. It dilutes connection. It erodes vitality.

More often than I care to admit, I get so stuck in my own head that I'm oblivious to what is going on around me. Countless times I've found myself walking around with my umbrella up long after the sun has come out. I have walked into the shower several times with my clothes on or searched for my sunglasses when they were sitting on my head. Like the figures in Bruegel's painting, I could all too easily walk by without flinching when Icarus came crashing down from the sky.

My patient Miranda told me about how she caught herself pouring salt on her food before she had taken a bite. "I think it is a metaphor for my life," she observed. During one session, we discussed mindless behaviors, and she increasingly became aware of the many "knee-jerk" reactions in her daily life. Without thinking, she emptied two packets of sugar into her coffee every morning. She reflexively turned on the television whenever she got into bed. She routinely checked her email the moment she woke up. She would reach for her phone whenever she was standing in line. Each time we met she would relate her latest "discovery" of such behaviors.

Miranda came to see me originally because of conflicts she had with her girlfriend. Whenever they argued, Miranda would retreat into herself and give her partner the silent treatment for twenty-four hours. This pattern would only intensify the original conflict. Rather than finding a mutually satisfying solution, Miranda's emphasis shifted to assigning blame and succumbing to self-pity.

Miranda recognized her behavior likely began when she was a child. Whenever her parents would fight, she would hide in her room, plugging her ears until the yelling stopped. This strategy served her well as a child but was undermining her current relationship.

To help Miranda disrupt the repetition of an old, negative pattern, I asked her to pay attention to the unique circumstances of a given conflict. So, for example, when an argument ensued about evening plans—her girlfriend wanted to stay in but she wanted to go out—Miranda was able to override her inclination to feel rejected. It wasn't that her girlfriend didn't care about her; Miranda's partner was exhausted from a work trip and just didn't feel like socializing.

"Sometimes it's nice *not* to be me," Miranda told me. It was liberating to let go of sulking, which she had assumed was just part of who she was. She compared the process to a snake's shedding its skin.

Ellen Langer has been studying mindful behavior for over forty years at Harvard. Her research shows that couples who resist dialed-in responses to each other have more satisfying and fulfilling relationships. Why? Because they resist becoming entrenched in routine. Because they don't jump to conclusions. Because they don't have predetermined expectations of the other person. These couples make a point of staying in discovery mode. They search for what's different about their mate rather than assuming that everything is known. They also avoid making predictions about how a situation will unfold. These couples view their relationship and each other as ever-changing.

"It is often assumed that a relationship stays stable when a person 'gets to know' their spouse. But once an opinion is formed, there is a danger that little opportunity is left for reevaluating, rediscovering that same individual," explained Langer and Leslie Burpee in a research paper on marital satisfaction.

If you want your relationship to stay fresh and to optimize interactions, make a habit of noticing what's new about your partner. Make rediscovery a daily practice. Resist the temptation to believe you know the person inside and out. Bringing an expectation of "not knowing" to all your interactions will increase your engagement with each other and enhance the time you spend together.

Put the Phone Down

"Did you smile or laugh a lot yesterday?" is one of the questions Gallup researchers asked people in order to generate its annual World Happiness Report. Finding humor in everyday life relieves stress and brings about feelings of fulfillment and vitality. A sense of humor can even change the way stress is perceived. The more humor in your life, the less stressful potential stressors turn out to be.

Laughing is primarily a social activity. For something to be genuinely funny, we have to give it our full attention. An amusing story a friend tells you doesn't get the belly laugh it deserves if you're checking your email at the same time. When we're not fully present we get less out of the moments we spend with others. Spending less time with our phones and more time with other people and without our phones is one way to increase the amount of laughter in our lives. If inattention diminishes the quality of our social interactions, then smartphones are the ultimate vampires of vitality. In one study, researchers asked three hundred people to go out to dinner with friends and family. Afterward, the diners were asked a number of questions, including how much they had enjoyed the experience. When phones were present, they enjoyed the experience less. Researchers had theorized that phones enabled people to entertain themselves when there was a lull in the conversation. In fact, the availability of a phone increased boredom.

Senior researcher Elizabeth Dunn concluded, "This study tells us that, if you really need your phone, it's not going to kill you to use it. But there is a real and detectable benefit from putting your phone away when you're spending time with friends and family."

A friend with college-age kids tells me the key to minimizing phone use during meals is for the entire family to pile their phones on top of one another—to "brick it"—at the end of the table. The first person who reaches for their phone either has to clean up or gets stuck with the bill. At home we now have a strict "no phones at the table" policy. It applies to everyone—my husband, the kids, and guests. Our friends are aware of this rule, and my kids often alert new guests. When it becomes a norm, it's easier to follow.

In 2018 Daniel Humm, chef and owner of the celebrated restaurant Eleven Madison Park, began offering diners little wooden boxes in which to place their cell phones for the duration of the meal. The point is to encourage guests "to enjoy the company of those at the table and be just a bit more present with one another."

Conversations can seem more superficial and less fulfilling when we are in a state of "absent presence," or the "split consciousness created by mobile technologies." Given the fact that 89 percent of smartphone users admitted to using their phone during their most recent social gathering, odds are they are compromising the quality of in-person interactions and consequently experiencing fewer uplifts in their everyday lives.

The Sharing Effect

In-person interactions are a reservoir of potential uplifts in our daily lives. How often are we leaving them untapped? Sharing a positive experience with another person makes it even more gratifying. But for the "sharing effect" to take hold, everyone involved needs to be fully engaged. Merely being in the same room or even sitting on the same sofa is not enough.

Participants in a Yale University study judged chocolate to taste better when they tasted it at the same time as another person. Researcher Erica Boothby explains: "We text friends while at a party, check our Twitter feed while out to dinner, and play Sudoku while watching TV with family—without meaning to, we are *unsharing* [italics added] experiences with the people around us."

Your cell phone may put the world at your fingertips, but it may also be making those closest to you feel as if you are a million miles away. Romantic relationships are particularly vulnerable. Philosopher Alain de Botton observed, "The constant challenge of modern relationships: how to prove more interesting than the other's smartphone." "Partner phone snubbing" or "phubbing" describes the habit of getting lost in your cell phone while in the company of your significant other. It is no surprise that phubbing is toxic for a person's love life—it leads to conflict, lower relationship satisfaction, and ultimately, unhappiness.

If you answer yes to any of the following questions, you may be unintentionally sending your partner a message that you value your phone more than you value him or her:

- Do you place your cell phone where you can see it whenever you are together?
- Do you keep your cell phone in your hand when you are with your partner?
- Do you glance at your cell phone when you are talking to your partner?
- Whenever there is a lull in the conversation, do you check your cell phone?

Even seemingly minor distractions erode the quality of an interaction. "Sorry, I was just checking my messages" might also be a message itself; one that indicates to your companion, "Sorry, my phone is more interesting than you are."

People who turn warmly toward a partner when their attention is summoned and express interest are more likely to stay together. More likely to separate are those couples who cannot be bothered to look up from a screen, or who keep playing Candy Crush and respond with hostility, "Can't you see I'm in the middle of something?"

Responding to a comment on Instagram, returning a text message, or updating Facebook may feel urgent, but it is not as important as the person you are with. Make the choice to put your phone away when you are with others, be it while having dinner, driving somewhere, watching a movie, or going for a walk. Whenever I'm driving anywhere with my kids, I ask them to put away their phones. The conversations during these drives often turn out to be good ones.

PART FOUR

Challenge Yourself and Embody Vitality

Chapter 16

CONSTRUCTIVE NEGATIVITY

I don't know what's wrong with me," my patient Daphne divulged.

Daphne had sought help following the loss of her ninety-two-year-old grandmother a month earlier. "I know I shouldn't be upset anymore," she said. "I should be happy my grandmother had a good long life, but I'm having a hard time moving on. My friend told me that sometimes people have abnormal reactions to loss. Maybe that's my problem."

I asked Daphne what "moving on" would mean for her. She explained that while she felt better than in the days just after her grandmother's death, she wasn't "totally back to normal." She still teared up at random times. When she was able to enjoy herself, she felt guilty afterward. She confessed that she wanted to talk about her grandmother "probably more than is healthy." Once, Daphne picked up the phone to call her grandmother to share a funny story. Another night, flipping through TV channels, she clicked past *I Love Lucy* and thought, *That's Grandma's favorite.* No, she reminded herself. It *was* Grandma's favorite. Daphne recalled how much she used to love watching the show with her grandmother. *Lucy* was too silly for her, but she loved how her Grandma would howl with laughter beside her. As we've seen, there's something magnetic about watching someone you care about laugh. They draw you into their joy, and the shared experience enhances the moment and deepens the connection.

Now these once happy memories filled Daphne with sadness. She was still going to work, talking to friends, running errands, doing laundry—keeping up with life. If she had been staying in bed all day, isolating from friends and not performing her basic daily activities, I would have been concerned that she was experiencing a more complicated grieving process. People with PCBD (persistent complex bereavement disorder) typically feel intense sorrow and a longing for the deceased that does not lessen over time. They become preoccupied with the person who died and sometimes express an urge to join them. They refuse to accept the reality of the loss and have trouble reminiscing about their loved one positively. Often they react irrationally, blaming themselves and believing that they let the person down.

Whenever doctors assess a patient, we generate a list of what could be causing the current symptoms. Obtaining the patient's full history, ordering lab tests, and performing a physical exam are all tools that help narrow down the list. This process of crossing potential diagnoses off the list is known as "ruling out." Given Daphne's presentation, it was easy to rule out PCBD. What she was experiencing was normal grief, a natural response to the loss of someone she loved dearly.

Everyone adjusts differently to loss. Being patient with yourself can help you adapt to the new normal. I told Daphne that replacing the present tense with the past tense can take a while, but it's also a perfectly healthy response. I shared a quote from British psychiatrist Colin Murray Parkes, who wrote, "The pain of grief is just as much a part of life as the joy of love; it is, perhaps, the price we pay for love."

What did concern me about Daphne's emotional state was not her grief but her guilt about that grief. As a society, we are constantly besieged with messages telling us to be happy. Feeling anything less than upbeat is considered abnormal, a problem that must be fixed. We have become increasingly intolerant of negative feelings: If you're sad, cheer up. If you're anxious, chill out. If you're angry, let it go.

Increasingly I meet patients who are convinced that bad feelings are unacceptable and should be medicated, controlled, or pushed aside. Friends tell them to focus on the good and count their blessings. Thinking happy thoughts *can* be a useful strategy, especially if you're in a situation where you lack control. For instance, if your flight is delayed, it might be helpful to reframe the extra time you've gained in the airport as an opportunity to read a book and catch up on email. But in situations where you do have control, putting a positive spin on a bad situation won't help. For instance, if chronic lateness leads you to get a poor performance review, then telling yourself that everything will be all right won't help you keep your job. On those occasions, problems need to be solved not by shifting your attitude but by actively making a change—like setting your alarm for half an hour earlier.

No doubt psychiatrists have contributed to a don't-worry-be-happy mantra with our smorgasbord of diagnoses and overzealous prescription pads. So have the champions of positivity who insist that happiness is a choice and every loss has a silver lining.

I strongly believe negative emotions are worth listening to. Feeling bad is part of being human.

Discomfort Is Data

One myth about emotionally healthy individuals is that they don't get sad or angry. Or if they do, they've learned how to grin and bear it. When someone cuts these people off on the highway, they smile. When their boss gives them a new project on a Friday afternoon that's due on Monday, they respond, "No problem!" Although suppressing emotions is effective at stifling a potentially harmful impulsive reaction like punching the wall or getting into a fight over a parking space, it's actually not a healthy long-term strategy for managing negative emotions. Habitual suppression comes at a real cost, most likely increasing the risk of dying from heart disease and even certain forms of cancer. Nor is it good for mental health. Suppressors are also more likely to be depressed and to lack social support.

Emotionally healthy people don't avoid negative feelings. They accept these emotions as a normal part of life and use them as valuable information. A certain amount of emotional discomfort alerts us that something isn't quite right and requires attention and possibly action. When used effectively, negative emotions can prompt us to change our behavior and help us to guide a situation in a new direction. (Think Miss Clavel from the children's book *Madeline* who, in the middle of the night, turns on a light and declares, "Something is not right." This feeling prompts Miss Clavel to race to the dorm, where she discovers that Madeline is in medical distress.)

Mike, age forty-one, came to see me after splitting up with his girlfriend. At the beginning of the first few sessions, before I was even able to say a word, he would jump in, filling me in on the details of the week: what he had done, how many miles he had run, whom he had had dinner with, and so on. He talked *at* me, not with me, snowing me with the minutiae of his life. When he did describe his breakup, he didn't seem particularly sad, even though it was the fourth relationship in five years that had gone sour. He was feeling frustrated but determined not to let the split bring him down. Mike didn't want to talk about his current emotional state. He wanted to focus on the future.

"What's the point of talking about it? What's done is done," he said. He portrayed himself as a master of moving on. "Isn't that healthy?" he asked. He had already gone on two dates. He had joined a cold-water swimming group and was training for the marathon. Unwilling to confront his sadness, he much preferred to keep moving, literally and emotionally.

Mike was correct to believe that ruminating can leave someone stuck in a relentless cycle of distress. But he was also a rationalizer, insisting, "She wasn't right for me in the first place." Rationalizing protected him from dealing with unpleasant emotions and feeling bad about himself, but it also prevented him from gaining any insight that would prevent his next relationship from repeating this same pattern.

Such self-protective measures may shield a person from facing uncomfortable truths, but they don't make those truths go away. Still, it's a common tactic for avoiding responsibility for one's own actions. A student gets a C on a paper and dismisses the bad grade as not mattering all that much. An employee receives negative feedback about a presentation and blames the client. An athlete falters in competition and convinces himself it was the referee's fault. "Full steam ahead!" is an effective strategy if you're a ship trying to reach maximum speed, but not if you're a human being trying to live a vital life.

Feeling disappointment better enables you to learn from your mistakes and also provides motivation to work harder the next time. In a study, "Emotions Know Best," participants were asked to complete a task to win a cash prize. While performing the task, the members of one group were told to focus on how they felt afterward. The members of the other half were told to rationalize why they didn't succeed if they lost. The task was rigged so that all the participants failed. When asked to complete a second task, the group that allowed itself to feel disappointment exerted 25 percent more effort than the rationalizers.

When it came to emotions, Mike was like a kickboxing champion, masterfully deflecting any attempt to talk about his feelings—using a jab of humor here . . . a rationalizing hook there . . . and finishing his opponent (namely, me) off with an uppercut of distraction. Pressing him to talk about uncomfortable feelings only shut him down more. During any lull in the conversation, Mike would crack a joke or start talking about the Knicks. During our third session, I asked him why silence made him so uncomfortable. He explained, "Well, that means you will want to go deep, you know, have a real conversation."

I asked him to consider why he had come to see me in the first place and continued to show up each week. Didn't that mean that part of him was at least curious about how a real conversation might unfold? I told Mike bluntly that he was wasting his time and money otherwise. It was a risk for me to push in this manner, but Mike heard my words. He acknowledged how his tendency to "keep it light" had become a recurring issue in his relationships and had contributed to the breakup with his last girlfriend. She told him he was exhausting to be around, always craving her attention but refusing to connect on a deeper level. He admitted how much he hated being alone and how being around others was more about distraction than genuine connection. I pointed out that this was exactly what he was seeking in our sessions, too.

For as long as Mike could remember, ignoring uncomfortable emotions was his preferred way of dealing with them. I asked him if he ever remembered getting upset as a child. "Not really," he replied. He couldn't recall any tantrums or meltdowns. "What did you do when you got angry or felt sad?" I asked. "I just kept going, no matter what. I would watch TV or hang out with a friend or play video games or soccer." When I asked if there had ever been something that keeping his mind or body busy hadn't solved, Mike then told me about his older brother, Philip. He had died in a car accident when Mike was seven. His brother had been sixteen at the time and driving alone. Mike recalled overhearing hushed conversations about alcohol and speed but didn't know much about the details surrounding the crash. His family dealt with the unfathomable loss by not talking about it and staying busy. Whenever people would ask his mother how she was doing, her reply was always the same: "Just keeping busy." This response taught Mike not to raise the topic or even bring up his brother's name, knowing it might upset his mother. He immediately made the connection between his mother's response to pain and his own.

Expand Your Vocabulary

Understanding why he had so assiduously avoided discomfort was an interesting insight for Mike, but he cared more about learning how to become more comfortable with feeling uncomfortable. Expanding his emotional vocabulary was a good place to start. When something was bothering him, I pressed him to be more precise about what he was feeling, instead of habitually rationalizing or avoiding. Labeling his emotions helped him feel more empowered to deal with them. For instance, during one of our next sessions, Mike mentioned his ex-girlfriend had sent him a text to ask if he could leave the pair of sneakers she had left at his house with her super so she could pick them up.

"It's just annoying," he remarked.

I encouraged him to be more specific. "The text stings because it feels so final," he explained. "Having those sneakers in the closet made me feel connected to her. They always gave me a little bit of hope that we would get back together. I have been hanging on to this fantasy that we would meet up for coffee so I could return the sneakers. Then we would have this great conversation, and I would tell her that I'm in therapy and working on my issues. I hoped she would agree to give our relationship another chance. This text extinguishes all of that." By putting his feelings into words, he was able to articulate precisely what he hoped for. It also gave him the courage to text her to suggest they meet for coffee.

The next time you find yourself feeling down, be as specific about the reasons for your state of mind as possible. Do you feel frustrated? Disheartened? Despondent? Exasperated? Instead of resigning yourself to a generalized negative mood, try to identify your emotions. Distressing feelings are less likely to dominate your attention and dictate how you behave if you can label them.

Pinpointing what it is that's upsetting you empowers you to seek a solution and to tailor a response. For example, recognizing that you are feeling dismissed by a colleague might prompt you to speak to the manager or perhaps go for a walk outside. Simply feeling "bad" about work doesn't provide you with any useful or actionable information, but hovers over you like a cloud and can easily rain on other aspects of your life. Nebulous feelings might manifest later as irritation with your partner or impatience with your child. Identifying what upset you and putting a name on the emotion is like sealing off a crime scene with police tape.

A clearly demarcated problem is less likely to become an emotional boomerang.

We're taught to consider emotions and moods as binary: you're happy or sad, calm or anxious. Ask your friends, "How was your day?" or "How are you?" and they'll probably answer along positive or negative lines. There is value in recognizing that far more nuance exists in emotional states than we often allow for, and that negative and positive emotions can coexist. Evidence suggests that people who are able to recognize and experience positive emotions alongside negative ones are more resilient and better equipped to handle adversity. They also might be happier.

Feeling a full range of emotions is more important than being positive all the time. A study concluded that people who experience a wide mix of emotions—emodiversity—had better physical health and were less likely to become depressed than those who are upbeat all the time. Rather than aiming to eliminate negative emotional experiences entirely, learning to allow and accommodate an array of emotions can help you experience the fullness of life.

Thinking Makes It So

People who have a negative view of negative emotions tend to be more negatively impacted by them. A bad mood doesn't last as long nor is it as physiologically distressing to those who are able to accept and find value in it. A "good cry" isn't a paradox. Shedding a tear can be a useful way to relieve emotions and even boost your mood. People who believe crying makes them seem pathetic or weak can't access this relief. Attitude and perception make a difference in pretty much every corner of our lives. Health researchers at Penn State found that no matter how many or few potentially stressful events a person faced, individuals who didn't *perceive* the events as stressful felt better. These people also displayed higher heart-rate variability, an indicator of resilience, than those who perceived the events negatively. Simply put, it is not the traffic jam that matters, it is how stressful you *perceive* the situation to be. As Shakespeare said, "There is nothing either good or bad but thinking makes it so."

Sara Blakely, the founder of Spanx, recalls that when she came home from school, her father would invariably ask, "What did you fail at today?" Most American parents would worry that such a question might scar a child for life. In a culture that demands perfection, talk of failure makes people uncomfortable. I think that Blakely's father's question has merits, his point being that, if you weren't failing, you weren't trying. From his perspective, setbacks and the accompanying disappointment and frustration were a positive sign and something to be proud of. Blakely credits her success, in part, to embracing this attitude. Being unafraid to face the discomfort of failure made it easier to press on. It also made her more resilient in the face of uncertainty.

The need for certainty drives a great deal of what we do on a daily basis. It is human nature to want to be "in the know." The world feels safer when we believe we can predict what's going to happen next. Not knowing how something is going to turn out can be extremely uncomfortable, especially for those with anxiety. The daughter of one of my patients accepted an early offer from a college she had no interest in attending just so she could put the process behind her. Knowing where she would be going in September mattered more than enduring the torture of waiting until March to find out. Unsurprisingly, a study showed that when faced with the choice of getting an electric shock now or in the future, most people would choose to get it out of the way as soon as possible. Some even said they would accept more severe pain to avoid having to put it off.

Some people will take any steps to avoid the unknown. Steering clear of unfamiliar situations or outcomes they cannot control, constantly needing reassurance, rushing decisions, or becoming paralyzed with indecision are among the many ways fear of the unknown can manifest in our lives. One patient of mine was six weeks into a new relationship.

Mauricio really liked his new girlfriend and liked her friends, too. It was one of the few times he said he felt comfortable with a woman, just sitting quietly and not needing to make small talk. But a few nights earlier she had sat him down and told him she wanted to know where the relationship was going. He answered that he felt it was going really well, but instead of reassuring her, his response upset her. She wanted specifics—did he see them becoming more serious? Could he envision getting engaged before the end of the year? She wanted definitive answers, but Mauricio didn't have them. His plan was for them to spend more time together and see where things went. Her need for certainty became a central issue in the relationship and, ultimately, the reason they broke up.

What frustrated Mauricio most was that even if he had told her what she had wanted to hear, it would have provided illusory assurance at best. "Who knows what might have happened?" he said. "After six months, she might have gotten sick of me and left me heartbroken. We'll never know." Like the people in the study who preferred to experience pain sooner than later, for his girlfriend, waiting itself was intolerable.

Find the Null Hypothesis

We would all like definitive answers to important questions and guarantees that our decisions are sound. But a relentless need for certitude can interfere with the possibility of expanding beyond the invisible cages in which we sometimes confine ourselves. When I'm advising patients on how to improve their tolerance for ambiguity, I ask them to apply the scientific method. When a scientist performs an experiment, she will use observations and knowledge to formulate a hypothesis, an educated guess, about what she thinks will happen. Accompanying every hypothesis is a null hypothesis—the assumption that the opposite of the predicted outcome might be true. For instance, if my hypothesis is that sunlight increases the rate of growth of the avocado plant on my windowsill, then the null hypothesis is that the rate of growth is unaffected by sunlight.

When a patient is having a hard time managing uncertainty or is feeling overwhelmed, I encourage them to formulate their hypothesis about the situation but also to consider the null hypothesis. For example, I asked Mike to explain his hypothesis that "going deep" was a waste of time. He quickly offered a number of reasons: "Who cares?" "The past is past." "There is no point." Then I asked him to consider the null hypothesis. What if it's *not* a waste of time? What if the opposite is true? Having Mike ask himself these questions enabled him to imagine an alternative to his preconceived ideas and projected outcomes.

We all have a tendency to jump from hypothesis to conclusion. We have ideas about ourselves and the world that are so deeply entrenched that we no longer notice them. They are our truths, our general rules about the world and our place in it. Here are a few examples that I've heard repeatedly over the years:

Relying on others is a sign of weakness.
Being successful is what matters most.
Nothing is worse than failure.
Nobody can ever tell me what to do.
I need people to like me.
People will always disappoint me.

Psychologist Karen Reivich calls these firmly held convictions "iceberg beliefs," because they are frozen, lurking beneath the surface of our lives. Sometimes we see the tips of our particular icebergs when we find ourselves overreacting to a situation or losing our temper because of something that was merely a minor irritation. Icebergs typically cluster around three main themes—acceptance, control, and achievement. To identify your personal "icebergs," consider what little thing typically sets you off. Is it a disruption to your schedule? Is it feeling unappreciated? Is it a need for perfection in yourself or others? Icebergs house our deepest insecurities and self-doubt and are often activated when we're faced with uncertainty.

A few months into treatment Mike arrived at a session having listened to a podcast with the poet David Whyte that he felt captured his own perspective. He quoted Whyte: "I began to realize that my identity depended not upon any beliefs I had, inherited beliefs or manufactured beliefs, but my identity actually depended on how much attention I was paying to things that were other than myself."

People sometimes go to therapy with the hope of eradicating all worry, uncertainty, and darkness from their lives. They see advertisements for medication on television that promise a life of unbridled joy. As if they were witnessing cosmetic surgery for the soul, they first see a "before" picture of a person looking miserable, worried, and alone, and then an "after" picture of the same person dancing the night away without a care in the world. In my experience the best psychiatrists prescribe medication when indicated but never promise it will whitewash discomfort or erase distress. They teach patients how to function well even on tough days, which becomes possible only by accepting negative emotions and not running from them. "Don't Worry, Be Happy" is a fine song (though way too overplayed, if you really want my opinion), but that philosophy can hold you back. "Worry, Explore Your Feelings, Name Them Precisely, Take Appropriate Action and Happiness Might Follow" is not as catchy a song title, but it's a better prescription.

If you are in a funk or particularly bad mood, channel your inner Sherlock Holmes and ask yourself, "What can I learn from this?" Find the clues that lead you to determine what triggered the mood. Is there something else going on that you need to address? Most important, don't beat yourself up for being in a bad mood. The truth is that occasional bad moods can be part of a good life.

Feeding Beasts

Some of our most uncomfortable negative emotions are directed toward others. Doctors know this firsthand. Most of us have great affection for our patients, but the uncomfortable truth is that there are some we see on our daily schedule and dread the session. During training I had one patient who was—how do I put this?—not my favorite. Between our weekly sessions he'd constantly leave urgent phone messages. "It's Zack. Call me back—it's important." No "Hello, Dr. Boardman." No "Would you mind?" No "Please." At first I would get right back to him, assuming he did need the help of a medical professional. With time, however, that proved not to be the case. On one occasion the "important" issue Zack needed to share was that he was thinking of visiting his parents over the holidays. This call came in the middle of July.

As the messages kept coming and coming, I finally discussed Zack with my supervisor. "When you listen to all telephone messages that patients leave for you, which patient do you call back first?" she asked. "Which patient do you call back last?" I admitted that I called the true emergencies back first, and then the "easy" patients—the ones who were pleasant and appreciative. Zack was usually the last patient on my callback list. "That should tell you something," my supervisor observed. It was likely that other people in his life had similar reactions to Zack. His unquenchable need for attention pushed people away. He was probably the last person on their callback list, too. My negative reaction to him didn't mean, though, that I should refer him to a new psychiatrist. It was instead valuable clinical information.

During our next session, I decided to address the issue of the "important" messages head on. This opened up a discussion about his neediness. Zack recounted that he had broken up with a girlfriend because he didn't feel "fully loved." Whatever she did to express her love for him, it was never enough. She told him point-blank that he was too high-maintenance.

"I just can't feed the beast," she explained, referring to his bottomless need for attention.

It became clear that Zack's demands for my own attention were a test to see if I genuinely cared about him. Instead of holding them against him, I realized they represented a fear of being forgotten and unloved. Learning to use my negative responses to certain patients as revealing clinical data has helped me better understand them. I know if I feel bored in a session, it's not because the patient is boring, it's because he's not talking about what's really on his mind. As much as I listen to everything that patients tell me, I am also trained to listen to what they *don't* say. Which stories are left out? What details are glossed over? Sometimes the reason given for coming to see me isn't why a person is really there. It is not deception, just unexplored terrain. The fact that a person is having difficulty with their teenage daughter might be what gets them in the door, but a conflicted relationship with their own mother and the fear of reliving that relationship is what troubles them deep down.

The dynamic works both ways. Frustration or irritation with one's psychiatrist isn't necessarily a reason to quit treatment. It can also mean that you're making headway. Exploring issues that make you feel uncomfortable can be painful, but it can also be a path to positive change.

Delicious Ambiguity

Professional figure skaters are awe-inspiring athletes not only because of their physical grace but also because of their psychological strength. Their spins, twirls, and leaps defy everything that the brain would naturally instruct a human to do. If I feel myself slipping backward on a slippery surface, my natural instinct is to shift my weight forward. As my entire upper body threatens to become parallel with the ground, my arms will stretch out in front of me. This is an automatic reflex designed to protect me. In fact, if you ever spot me on a skating rink, that's pretty much how

I'll look. But figure skaters embody the triumph of will and practice over instinct. They learn to spin with their heads leaning backward, which requires overriding vital protective reflexes. Still, if an ice skater is walking down the street and trips, he will do what all of us do—put his hands out to brace for a fall. Figure skaters adapt their responses to the context they're in. This ability to rewire circuits in their cerebellum is relevant to how we deal with negative emotions and hassles in our everyday lives.

Most of us have a go-to coping style to help us manage discomfort and uncertainty. Do you suppress your feelings? Distract yourself? Rationalize? Reframe? Do you ruminate? Reflect? Depending on the particular context, all these responses can be useful. If you feel like shouting at the driver who cut you off, it's probably better to bite your tongue and suppress your feelings. If you're having an MRI, distraction can be a godsend. Even ruminating can be helpful if it is task oriented and focused on a specific situation you can rectify.

It is helpful to adapt your response to what best suits the situation. I had a college-age patient who ruminated about situations beyond her control—a well-established dead end. Whenever she took an exam in college, all she could think about afterward were the questions that she might have gotten wrong. And when she didn't have something in the past to fret about, she'd ruminate about the future, imagining potential disappointments and missteps as she tried to fall asleep. With practice she learned to override this tendency and to accept uncertainty. She deployed such strategies as self-distancing and spending time in nature. Like an Olympic figure skater twirling with an arched back, she triumphed over her reflexes. The words of comedian Gilda Radner resonated deeply with her: "Now I've learned, the hard way, that some poems don't rhyme, and some stories don't have a clear beginning, middle, and end. Life is about not knowing, having to change, taking the moment and making the best of it, without knowing what's going to happen next. Delicious ambiguity."

Chapter 17

EXPAND YOURSELF

"Do you have a hobby?" I asked the twenty-two-year-old woman interviewing for an assistant position in my office.

She looked at me as if I was from another century. "Are you asking if I collect stamps or something?" she responded quizzically.

I explained that I was curious about what she did in her spare time. She said that while at college she had been involved in a number of clubs and organizations. She wrote for the school paper, provided after-school tutoring for inner-city kids, and played club soccer. But since graduation and beginning a full-time job as an assistant at a PR firm, she was either at work or trying to chill out. I pressed her on her conception of "chilling out," which she described as looking at her phone, her computer, or her television.

I have seen this pattern in the lives of many of my patients. Passive leisure becomes the default with the common explanation, "I don't have time for anything else." Surveys of leisure time suggest that even busy people have more free time than they might think—close to four hours during weekdays and more than five hours during weekend days. The biggest chunk of free time is spent in front of a television or a screen. Screens trump socializing, exercising, and pretty much everything else. The average twenty-five to thirty-year-old spends fewer than nine minutes a day reading for personal interest. Physical activity is also low on the list of chosen leisure activities. Thirty-five to forty-four-year-olds spend fewer than twenty minutes a day moving their bodies.

Given the choice, we opt for mindless hedonism. A study, "The Paradox of Happiness," explores why we gravitate toward devitalizing pastimes like watching TV, surfing the internet, and checking social media, even though most of us recognize that these activities give us a fleeting boost at best. This hedonic approach maximizes pleasure and minimizes discomfort, but it doesn't lead to vitality. "Because work is often highly demanding, people feel their free time is too precious to risk losing to yet more challenging activities, so they resign themselves to experiencing happiness through easy 'relaxing' entertainment," explain the researchers. These activities even have a name. They are called "demand shielding" because they require little of us physically, intellectually, and socially.

But the relief provided by less-challenging activities is temporary. The moment we return to reality, the stress comes rushing back. It is more beneficial to adopt what is known as "a eudaemonic approach," which includes activities that provide the sense you are living your life in a full and satisfying way. Effortful intentional activities that stretch the body or the mind may not be experienced as pleasurable in the moment, but when we recall them, we're more likely to think, *Wow, that was awesome!* Or, *That was time well spent.*

Oxygenate Your Mind

If kicking your feet up doesn't help you recharge, what does? It might not *sound* relaxing, but try engaging in activities that build psychological resources, like competence and self-efficacy. Learning something novel is a good example. One study found that mastering a new skill or challenging yourself intellectually is more effective at reducing anxiety and building resilience than demand-shielding activities. Using our minds feeds our psychological need for growth, discovery, and expansion. In addition, gaining new skills or knowledge directs our focus outside of ourselves. Astronomy professor Abraham Loeb once observed, "Learning means giving a higher priority to the world around you than to yourself."

When you're feeling stressed, instead of retreating into yourself, raise a periscope and look at the world—and yourself—through a broader lens. When you engage in interesting activities and experiences, you expand your sense of self, which in turn increases the effort you're willing to exert in other domains.

During one study, participants were given a list of eleven facts. Half of the group was provided with facts that were "high expansion," such as, "Butterflies taste with their feet." The other half was supplied with a list of facts that were common knowledge, such as, "Butterflies begin life as a caterpillar." Both groups were then presented with a challenge to solve tricky puzzles. The group that was exposed to the more interesting information was more motivated to find solutions than those provided with run-of-the-mill facts.

Expanding experiences energize us. To assess daily positive experiences and well-being, one of the questions on a Gallup poll conducted in 2018 asked: "Did you learn or do something interesting yesterday?" I hope you will be able to answer with a resounding yes.

Effort Begets Effort

Folding in on ourselves, sticking to mundane routines, and turning our backs on new challenges or novel experiences limits possibilities and narrows perspective. To paraphrase Winston Churchill, more than intelligence or physical strength, the key to unlocking potential is effort, and the best way to unlock effort is to participate in activities that broaden and stimulate you.

To determine if an activity will be self-expanding, try asking yourself the following questions:

Will the activity increase my knowledge or result in me learning something new?
Will the activity broaden my perspective or awareness?
Will the activity increase my ability to accomplish new things?

Functional imaging reveals that, in addition to building self-efficacy, engaging in expanding activities like solving puzzles and playing challenging games activates reward centers in the brain. It is possible that taking part in these types of activities can also help people quit smoking and reduce other addictive behaviors.

Engaging in actions that stretch one's brain or body—or both—is rewarding. It feels good, not in a superficial cotton candy way, but in a kale salad way. Keeping this in mind, instead of watching *Frozen* for the tenth time with your kids this weekend, teach them how to play a card game. Instead of checking Instagram first thing in the morning, learn a new word with a 365 page-a-day calendar. Instead of scrolling through Twitter before bed, crack open a biography or novel. It took me over two months to read *A Little Life* by Hanya Yanagihara, but the effort was far more gratifying than obsessing over breaking news.

Engaging in expanding activities also increases what is known as "self-complexity." People with low self-complexity view themselves in narrow terms. If you are a lawyer and pin your identity on your profession, a bad day at the office renders you less resilient to stress than if you see yourself as wearing many hats, such as mother, friend, yoga practitioner, volunteer, amateur astronomer, piano player, up-and-coming baker, fiction reader, and so on. Not placing all your eggs in one basket lessens the impact of negative experiences and promotes flexibility in how you respond to them.

You've Got to Go with the Flow

Expanding activities help us transcend ourselves and give rise to flow—an all-encompassing mental state that is simultaneously engaging and challenging. When you're in flow, you're so deeply immersed in what you're doing that you don't think about anything else. Ruminating and introspecting aren't an option. Flow is hard to describe, but you know it when you feel it. Obvious examples of flow occur in sports when an athlete finds the zone—that optimal state of experience when they deploy their skills to meet the challenge at hand on the soccer field, balance beam, or when running a marathon.

Flow activities require focused attention and significant effort, and yet the experience is more rewarding than draining. The climber climbs because she loves the act of climbing itself—not just as a means to reach the summit. I am convinced that one of the reasons spinning classes are so popular today is that they provide a flow experience. Studies show that most people cannot leave a mobile device alone for six minutes without checking it, but spinning enthusiasts set it aside for those sacrosanct forty-five minutes. The physical demands of spinning, combined with immersion, an inspiring teacher, and a great playlist, creates what many spinners describe as an almost sacred experience. As one spinning fan said, "As good as it is for my ass, it's better for my head. It's mental floss."

Flow occurs when we deploy our talents and mobilize our strengths. We feel energized, engaged, and soothed—essentially the opposite of feeling tired, stressed, and bored. Flow experiences facilitate well-being.

Where Skill and Challenge Collide

Artists often describe the creative process as an experience of complete absorption in which time melts away and the world fades into the background. "It's more like I'm having an experience than making a picture," painter Cy Twombly once explained.

Making music is another flow generator, as captured in Albert Einstein's letter to his eleven-year-old son: "I am very pleased that you find joy with the piano. . . . Mainly play the things on the piano which please you, even if the teacher does not assign those. That is the way to learn the most, that when you are doing something with such enjoyment that you don't notice that the time passes. I am sometimes so wrapped up in my work that I forget about the noon meal."

Einstein isn't channeling a Tiger Mom urging his son to play until his fingers bleed. Rather, he is counseling the boy about how to find the time-melting sweet spot where skill and challenge collide.

Flow isn't limited to athletes and geniuses. Mere mortals can experience it on an everyday basis while doing such simple tasks as washing dishes, driving on the open road, singing in a choir, getting lost in a good (or trashy) book, practicing your dance moves, or playing mah-jongg. Watching a play or attending a concert can also be a total immersive experience and offer a sense of flow.

Surgeons experience flow during operations, and lawyers can even get into it while reading contracts. I sometimes experience flow while writing or doing research or when seeing a patient. Flow may occur during a social interaction, when talking with a good friend or while playing with a baby or going for a walk with your partner.

Flow experiences are essential for everyday vitality. They lift us out of ourselves and replenish depleted resources. They cultivate resilience. When you're in flow, you're not thinking about the party you weren't invited to or what you're going to say next. Any tendency toward pettiness or self-focus evaporates. The more flow you experience in your daily life, the more vital and replenished you will feel.

Today, there are fewer and fewer opportunities to experience flow. Even if we are able to find them, we're likely to be constantly interrupted by texts and emails and notifications. When is the last time you spent an hour in flow without distraction? Putting flow back into everyday life requires effort, but the reward will be the experience.

ACTIVATE YOURSELF

How can you build more flow into your everyday life? Start by putting your phone away and spend twenty minutes focusing all your attention on one activity. No multitasking or having the TV on in the background. Cooking dinner, gardening, making art, writing a letter, playing an instrument, walking outdoors are all good examples. Can you ride a bike to get around? Even if it's just on the weekends, riding a bike is not just an activity for kids. Give it a try. It is better for the environment and for your head and your heart. Plus, being outdoors is an added benefit.

Although most of us are keenly aware of how good we feel when we engage in activities that require an investment of psychological or physical effort, we often put them off because they seem daunting. The key is to predecide, preplan, and prearrange flow-promoting activities to lower activation energy. As a reformed couch potato, I feel much better whenever I *do* something and override the temptation to "twiddle my thumbs," as my mother liked to say. I am not suggesting that you do away with passive leisure entirely. It can provide a much-needed break from the demands of daily life, but it's counterproductive when it becomes your default way to spend downtime. Don't confuse moments of pleasure with those of fulfillment. Mindless relaxation is not the same as meaningful engagement. If bingeing a television program restores you so that you can return to flow activities, by all means, put your feet up. But be aware that you're unlikely to find flow in mind-numbing activities. Feeling as if you've wasted your time or that you're not using your skills or challenging yourself can lead to apathy, which is about as far away from vitality as you can get. There is a lot you cannot control in your everyday life, but you can be deliberate about how you spend downtime and choose a hobby that oxygenates your mind and stimulates your soul.

Patients frequently come to me with a yearning to tunnel inward. I often recommend that looking outward and losing themselves in activities that stretch them is a vital component of personal growth.

GET CREATIVE

You may not think of yourself as a creative type, but doing something artistic can give your brain a break. Making art reduces cortisol, a widely studied marker of stress, even in people who have no experience doing so. One study had participants provide a saliva sample before they were invited to get creative by using markers, clay, and collage materials. The participants were told they were welcome to work with all the different media and they had free rein over what they did. No final artwork was expected. After forty-five minutes a second saliva sample was collected, and cortisol levels had dipped. Participants also reported feeling better afterward. They described the experience as giving them a sense of freedom, enjoyment, enthusiasm, and a feeling of losing themselves in the work. Most said they wanted to make more art in the future.

When it came to making art as a child, I was hopeless. I can't draw or paint. You don't want to read my poetry or hear me play piano. One summer I did teach myself how to needlepoint, which requires more concentration than imagination. I could meticulously follow a pattern, one stitch at a time, and completed some smaller projects—a belt, Christmas ornaments, a coaster. (In my childhood home, there could never be too many coasters.) Decades later, as an adult, I realized that I missed the sense of actual creation. I decided to tackle a more intricate project: a pillow decorated with a picture of a dog. Did I mention that the pillow required needlework not just on the front but on three sides? There was a separate canvas for the dog's curly tail and a bottom canvas with the paws. At least ten shades of brown and beige were required to achieve the subtle markings of the dog's coat. A tartan ribbon around the dog's neck demanded painstaking precision. Getting the tail right was a formidable task. The pillow took months to complete and is today one of my prized possessions. A needlepoint pillow that belongs in a grandmother's parlor in the 1950s exists far beyond the reach of my husband's contemporary taste, but this pillow occupies prime real estate in our bedroom.

For people who are obsessed with being productive at work, consider the finding that those who engage in creative hobbies are actually better at their jobs. Hobbies give us an opportunity to take off our work hats and channel our energy in a different direction. There is satisfaction in spending time doing something purely for the love of the game. Hobbies create a sense of mastery, control, and growth; restore a sense of order in the chaos of daily life; and provide a change in perspective. I puff up like a penguin whenever I finish a puzzle or work with my daughter to construct a building with LEGOs.

"Many scientists say that their hobbies provide them with crucial opportunities to relax, to find satisfaction in completing small, defined projects and, occasionally, to make the kinds of insightful leaps that propel science forward," observes Alex Clark, associate vice president of research at the University of Alberta in Edmonton, Canada.

This might explain why Nobel Prize winners are two-and-a-half times more likely to have an artistic hobby than the average scientist. Hobbies likely opened their minds and enhanced their scientific creativity.

Expand Your Range

In the past few decades, our culture has moved away from hobbies. "Do one thing well," advised Apple founder Steve Jobs. (He didn't follow his own advice. He was passionate about calligraphy and music and reading everything from Shakespeare to Plato.) Emphasis is placed on achieving success in a single domain, specializing in a particular area, and devoting all your time and energy to drilling down at work. Dabbling is frowned upon, and becoming sidetracked is a disaster. But research suggests that people often get their most creative ideas when they are not "on task." Physicists and professional writers reported that many of their best aha moments occurred when they were not on the job. Before you dive into a career and leave all your other interests behind, consider that some of the happiest and most successful people capitalize on their downtime by engaging in stimulating activities that stretch their minds and expand their sense of self. They become learn-it-alls, not know-it-alls.

Building learning and discovery into everyday experience offers ballast and fosters resilience. One of my favorite books is *The Once and Future King*, T. H. White's retelling of the legend of King Arthur. While instructing his pupil, Arthur, the wizard Merlin also instructs the reader:

> "The best thing for being sad," replied Merlin, beginning to puff and blow, "is to learn something. That is the only thing that never fails. You may grow old and trembling in your anatomies, you may lie awake at night listening to the disorder of your veins, you may miss your only love, you may see the world about you devastated by evil lunatics, or know your honour trampled in the sewers of baser minds. There is only one thing for it then—to learn. Learn why the world wags and what wags it. That is the only thing which the mind can never exhaust, never alienate, never be tortured by, never fear or distrust, and never dream of regretting. Learning is the thing for you. Look at what a lot of things there are to learn."

Chapter 18

EMBODIED HEALTH

Get Some Sleep

Unusual sleep patterns can be both a symptom and cause of illness. Sleeping too little without experiencing fatigue suggests a manic episode. Sleeping too much without feeling rested is an indicator of depression.

While working as an intern, I didn't get much sleep myself. In medical school we used to joke, "I'll sleep when I'm dead." I told myself that the dark circles under my eyes were proof of my commitment to work. This was twisted logic. Those dark circles were actually a sign of exhaustion. My patients probably (and justifiably) wondered how, if I couldn't manage to take care of myself, could I take care of them? A healthy appearance is actually a strategy for appearing competent. When looking for leaders, people prefer healthy-looking faces.

Sleep is an essential ingredient of everyday vitality. Just one night of poor sleep can throw you off and result in a more stressful day. Not getting enough rest can magnify a bad mood and induce negative feelings like anger and nervousness and also inhibit positive emotions. Studies also show that fatigue may make a person more vulnerable to "cognitive interference," the experience of intrusive, unwanted, distracting, and potentially ruminative thoughts. In other words, it's hard to concentrate and think clearly. (Like you needed a study to tell you that.) Cognitive interference reduces productivity and magnifies misery. Being sleep deprived often distorts our judgment, compromises our decision-making, and increases our risk of engaging in harmful behavior. High school students who slept fewer than six hours a night were more than three times more likely to consider or attempt suicide.

Not sleeping enough can also turn us into grumpy loners. A UC Berkeley study showed that the less sleep people get, the less they want to participate in social activities. Nor are these exhausted people as welcomed. In the succinct summation of Berkeley neuroscientist Matthew Walker, "If you haven't slept, other people perceive you as 'socially repulsive.'" Think about that term for a moment—*socially repulsive.* When we're walking zombies, people usually want to get away from us. And they should. Being sleep deprived can bring other people down as well. After watching a sixty-second clip of a sleep-deprived person, healthy volunteers reported feeling alienated and lonely. Sleep-deprived people are not only at risk of becoming socially isolated; the unlucky individuals with whom they interact may also end up feeling that way.

Not sleeping well can be both a cause and consequence of daily stress. A tough day full of hassles, frustration, and tension can make it hard to fall asleep that night. Unpleasant interactions cast a long shadow. Interacting with a rude or sarcastic coworker and feeling disrespected at the office can follow you home and keep you awake at night.

Sleeping between seven and eight hours is optimal, and yet the results of the world's largest sleep study stated that nearly a third of Americans are sleeping six hours or fewer. Shortchanging sleep typically leaves people more vulnerable to setbacks, more irritated by hassles, and with greater feelings of loneliness. Even if you get enough sleep, a third of the people with whom you interact don't, and that can affect your day significantly. When we're fatigued, our emotional processing is thrown off balance. We interpret neutral information in a negative way. So even though you thought you had a perfectly pleasant exchange with a coworker, if she is sleep deprived, she might disagree and think you don't like her.

Most people recognize that sleep makes them feel more effective the following day, but there are those who still don't or can't make sleep a priority. Unlike the many other scheduled parts of their day, bedtime happens when it happens. We are rigid about setting the time for our children to go to bed but do not do so for ourselves. We schedule workouts, doctors' appointments, and meetings because they are important. Why not designate a specific time to go to sleep? Setting an alarm to alert you to go to bed can reinforce how important it is. You may sleep so long and so well that it will eliminate the need to set an alarm to wake you up in the morning.

I also suggest you set a notification one hour before bedtime as a reminder to start powering down—your devices and yourself. This means no more email, no video games, no work, no paying bills, and nothing stimulating or anxiety-provoking from that point on. This type of designated power-down ritual sends a signal to your brain that it's time to transition from work into relaxation mode and better prepares the body for sleep.

Looking at your phone before bed makes it harder to get to sleep and is associated with poor sleep quality. Sixty-eight percent of phone owners sleep with their phone next to their bed despite studies that show screen use just before attempting to fall asleep is particularly problematic.

When I ask new patients if they sleep with their phone on or in their bed, they usually nod. When I ask if they have ever been awakened in the middle of the night by a phone call, text, or email, they nod again. They tell me their phone is the last thing they touch before they go to bed and the very first thing they touch in the morning. "But it's my alarm clock" is the typical response to my suggestion to leave their phone charging overnight in another room. This is easily solved by buying a small travel alarm clock to use instead, which will help prevent you from looking at your screen right before going to sleep. You're also less likely to get sucked immediately into scrolling the moment you wake up.

Removing the smartphone from the bedroom is essential. Charge it in another room—but not the bathroom. (I don't want anyone being tempted to check his device if nature calls in the middle of the night.)

"How many hours do you sleep?" is near the top of the list of questions that I ask patients during an initial consultation. Many say they get less than seven hours. This is often accompanied by the line, "I'm just not the kind of person who needs a lot of sleep." The problem is that when we're sleep deprived, we don't even *know* we're sleep deprived. We think we're fine and operating at full capacity, even though our cognitive abilities and reflexes are dulled. My typical response is to ask these patients to go to bed an hour earlier than their normal bedtime for just one week. Without fail they are astonished by the improvement in their mood and energy level. Greater clarity is a common benefit. One patient likened it to having cataract surgery. "You don't know how much your vision has deteriorated until you can see clearly," he explained.

It is a virtuous cycle—if you sleep well, the following day will likely have fewer conflicts, less stress, greater productivity, and more positive emotions and experiences, which in turn will make it easier to sleep well that night. It is no wonder that Professor Orfeu Buxton, director of Penn State's Sleep, Health and Society Collaboratory, cited sleep as a powerful source of resilience.

Get Moving

After I ask patients about how much they sleep, I often follow up by asking how many hours they sit at their desks and how much time they spend outdoors. These questions were not emphasized in my original training. If a patient was clinically depressed, the textbook response was to write a prescription for an antidepressant. If the patient did not show signs of improvement within a few weeks, the dosage was increased. The aim was to "medicate and ameliorate." The focus was above the neck.

Today, it's been shown that a thirty-minute walk three times a week can be as effective as medication in relieving symptoms of depression. Stressed-out university students who began exercising a few times a week reported wide improvements across a number of areas, including better eating and study habits, reduced smoking and drinking, and being more responsible about spending.

Not long ago I took a refresher course to prep for a board certification renewal exam in psychiatry and neurology. The renowned psychiatrist leading the class spoke at length about new medications in the pipeline and off-label uses of ones already on the market. He discussed treatments like transcranial magnetic stimulation, vagal nerve stimulation, and ketamine, which held promise for our patients. His approach, like my training had been, was based on the biomedical model that assumes all mental illnesses are biologically based and arise in the brain, independent of behavior and environment. I raised my hand and asked a question about lifestyle interventions, specifically, exercise as a worthwhile approach to not only reduce symptoms but also to promote well-being. He smiled, rolled his eyes, and said, "We're psychiatrists, not trainers." The audience laughed. He continued. "It could help but it's not a primary concern. *First things first.*"

A prevailing belief among many psychiatrists is that physical activity is a baton that cardiologists and internists pass to social workers and trainers. This mindset explains why psychiatrists are not typically interested in their patients' physical fitness.

Jackson was a patient who came to see me after his wife of twenty years kept complaining about his bad mood, telling him he was "difficult" and "irritable." Even he admitted that he was "not in a good place." Six months earlier he'd undergone knee-replacement surgery, which forced him to stop running. The operation went well, but the recovery had consequences that weren't being addressed. Throughout his life, exercise had played an important role in his mental health. Instead of a pill, I prescribed swimming. It helped, and he started feeling like himself again.

Four out of five patients being treated for mental illness at an outpatient clinic said exercise helped to improve their mood and reduce their anxiety. Yet over half admitted it was a topic rarely discussed by their doctors. Fortunately, this is changing as recent research has found that exercise can not only help to treat depression but can lower the risk of developing it. Studies have shown that a sedentary lifestyle almost doubled the chance of developing depression, compared with an active one. It turns out it's not a bad idea for psychiatrists to think like trainers.

Start Skipping

For managing the slings and arrows of everyday life, physical activity may be the last thing on your mind but may also be one of the best things for it. A study of more than a million people indicated that those who exercised had up to five fewer bad days a month (days in which they felt stressed, depressed, or emotionally tapped out). The boost did not require Olympic-level training. A little exercise went a long way—just thirty minutes to an hour, three to five times per week, was sufficient. Team sports and cycling showed the greatest benefits. Mindful-based techniques like yoga and tai chi also were effective.

While exploring exercise options for yourself, also try to encourage loved ones to try them, especially during periods of transition and stress. Consider these suggestions:

- Invite a friend who has suffered a loss to go for a walk.
- Send a teenager off to college with a yoga mat.
- Offer to be a gym buddy to a friend going through a divorce.
- Watch a friend's baby so she can do an online workout.

Even less strenuous activities like housework can boost mental health. The results of related research showed that regular, everyday actions, like walking to the store or carrying your groceries up the stairs, can markedly improve a person's mood. Using a mood-tracking app that also measured physical activity, researchers discovered that individuals who had been moving in the past fifteen minutes were in a better frame of mind than when they had been reclining or sitting down. The study participants who moved a lot also reported greater life satisfaction than did the couch potatoes.

Getting off the bus a stop before your regular stop, using the stairs instead of the elevator, taking the dog for an extra loop around the block, parking a little farther away from your destination, and going for a walk after dinner instead of collapsing on the couch are all small but effective ways to increase the amount of physical activity in your day. Whenever you're in an airport, skip the moving sidewalks and escalators. Use your legs and move your body whenever you can. There is a clear link between ambulating and feeling good. Take advantage of every opportunity to move even when you're not in the mood.

When we're in good spirits, we walk with a spring in our step. Literally. Comprehensive gait analysis has demonstrated that depressed people walk more slowly, swing their arms less, and have a posture that is more forward-leaning, compared with upbeat people, who walk in a more upright position and tend to bop up and down with their arms swinging. Just from observing someone's posture, it's possible to deduce what is going on in her mind. Without even looking at an individual's face, people can tell who is winning a professional tennis match simply by reading the players' body language.

Mood has been shown to affect how we move, but the opposite is also true: how we move affects how we feel. Participants in a study who were prompted to imitate a depressed style of walking—minimal arm movement and shoulders slumped forward—experienced worse moods and were more negatively focused than those who were instructed to imitate a happy walk. Walking with your head lowered and shoulders drooping brings you down.

A few years ago I learned this firsthand from my daughter. After a long day she begged me to skip beside her down the sidewalk. I was exhausted and not in the mood. Plus, what self-respecting psychiatrist wants to be seen skipping down the sidewalk? Still, her pleas won me over. She took my hand in hers and began bouncing down the block. I had no choice but to follow along. Within moments my self-consciousness slipped away, and before I knew it, I was smiling—I couldn't help it. Try it yourself. Start skipping and notice how your mood shifts instantaneously. It is virtually impossible not to feel joy and vitality while doing so.

How we sit matters, too. In one study upright sitters reported feeling greater self-esteem, enthusiasm, and excitement than the slouchers, who reported feeling more fearful, hostile, nervous, passive, dull, sluggish, and sleepy. The authors concluded, "Sitting upright may be a simple behavioral strategy to help build resilience." Good posture can even bolster the performance of anxious students. Students who slumped in their seats didn't perform as well on tests. In comparison, an empowered upright position helps to optimize focus and build confidence, especially in those who tend to engage in negative self-talk. The position of one's body influences how a person processes information. Freud never articulated a clinical rationale for having patients lie down on a sofa and famously said that he insisted patients recline because he couldn't stand being stared at all day long. There is evidence, however, that being in a supine position reduces defensiveness and helps us learn from mistakes. But if you're trying to get work done or master challenging material, reclining is not ideal. As my children know, one of my pet peeves is when they lie on their beds to do schoolwork. Unfortunately, presenting them with evidence as to why this isn't a good idea falls on deaf ears. Thankfully, my patients are more receptive.

I discuss with all of my patients how posture can shape everyday experience. A woman who came to see me a few years ago with mild symptoms of depression is now vigilant about both sitting and standing tall. She wears a posture bra and bought a device that sticks to her back and vibrates if she leans too far forward. Instead of adopting a default slouchy stroll, she channels a ballerina's grace, imagining a straight line running from the floor to the top of her head. When she is having a bad day, she says standing up straight provides a sense of control and power.

"*I'm* in charge of how I stand and move in the world, not my mood or my phone," she told me.

To sit up, stand up, and physically move from one place to another can help you keep it together in emotionally charged situations. Anne Kreamer, author of *It's Always Personal: Navigating Emotion in the New Workplace*, explains how one of the best strategies to restore control when feeling overwhelmed is to simply get up and change your environment:

> Say you're in a meeting and you're having a confrontation with somebody who isn't understanding you, and you're about ready to blow up, or you feel tears coming on: go get a drink of water. (It's about the moving, not the water.) The physical movement begins to reset your parasympathetic nerve system, and put you in a different mindset and allows you to gather yourself so you can go back in and try to tackle the conversation with a fresh perspective. One forgets they have the power to do that, and it's really hard to remind yourself. But there are very few situations, even if you're in a tense negotiation with your boss, where you can't say, "You know, I need a drink of water, do you mind if I go get one?"

The bottom line, as Kreamer says, is: "Do something that allows you to change your physical relationship at the moment with that conversation or argument." A walk to the water fountain or coffee machine might be enough to shift your perspective and reset your emotions and recharge. When you change your physical relationship to a situation, you change your emotional relationship to it, too.

How we move and carry ourselves in the world influences how we feel and perform in subtle but significant ways. Keep this in mind while you scroll through Instagram, check email, or follow breaking news on Twitter. Your neck bends forward and your body hunches over. Given the finding that the average American spends more than 3.5 hours a day on his or her phone, this ubiquitous bowed stance might be taking a toll on our collective mental health. Be deliberate about how you hold your body. It will help you be more deliberate in your everyday life.

Go Outside

"Nature is fuel for the soul," says Richard Ryan, a professor of psychology at the University of Rochester. "Often when we feel depleted, we reach for a cup of coffee, but research suggests a better way to get energized is to connect with nature."

Most of us recognize that being outdoors boosts our mood, and studies show it reduces stress. Just twenty minutes in nature can lower levels of cortisol, a stress-hormone marker. Still, spending time outdoors is rarely a goal we seek. We tend to leave the comforts of home only on the way to other destinations. A study of twelve thousand-plus Americans found that more than half spend five hours or fewer outside . . . *each week.* Before you insist, "Not me!" actually count how many hours you spend outside on an average day. Be honest. Maybe twenty minutes to walk the dog or take a child to school, maybe ten minutes to go to the grocery store? And remember, being in the car doesn't count. The truth is, we're indoors most of the day at work, and leisure time is increasingly indoor oriented and spent in front of screens.

One of the best parts of immersing oneself in nature is that it lifts us out of self-immersion. A ninety-minute walk in green space was found to reduce anxiety and rumination—that repetitive rehashing of self-focused negative thoughts. One hour and a half sounds like a big commitment, but we routinely spend that same amount of time to watch a movie. Brain-imaging scans suggest that being in nature actually changes the brain. Neural activity in the subgenual prefrontal cortex, a brain region typically active during rumination, decreased in participants who took a stroll in the park.

There is also evidence that being in nature makes us nicer to one another. When mothers and daughters walk in the park versus in a mall, they actually feel closer and get along better. The participants of an experimental study reported that nature was more fun, relaxing, and interesting than being indoors. Spending time outside is not always an option, but experiment with grabbing some fresh air when you might not otherwise. Does your company cafeteria have an outdoor patio where you can eat lunch? Can you walk around the block while making phone calls?

If I could, I would prescribe a daily half-hour walk outside to all those able to do so. Many psychiatrists still hesitate to prescribe lifestyle interventions, but I would argue they are just what the doctor *should* order.

Chapter 19

FORTIFY THE BODY, FORTIFY THE MIND

Researchers conducting a 2015 British study asked participants to complete a daily food diary and also record their mood and behavior for two weeks. On the days people ate more fruits and vegetables, they felt not only happier but also more engaged, curious, and creative than they did on the days they ate fewer amounts of these foods. I know this sounds as if your mom conducted the study, but people who ate more chips were also more likely to report being in a bad mood. Although this study does not prove that munching on carrots will provide a sense of purpose, or that an apple is your ticket to vitality, the results are a reminder of the bridge that connects physical and mental health.

Feeling mentally strong isn't confined to what is going on inside your head. Psychiatry may be considered an "above the neck" discipline, but mental health includes the entire body. How you move, sleep, and eat affects your mood, which in turn affects your interactions and experiences. Every breath, every bite, every step has the potential to shift how you perceive the world.

Most of us are aware that a typical Western diet high in fat, sugar, and processed foods negatively impacts physical health. Fewer are aware of its negative impact on how we feel. A person's ability to focus may falter after eating just one meal high in saturated fat. Four days of eating an unhealthy breakfast—like a chocolate milkshake and a breakfast sandwich—can lead to impaired learning and memory as well as to mood swings in people who are otherwise healthy and physically fit. Just one holiday week of overindulging can not only slow you down but bring you down. On the flip side, researchers found that depressed college students whose diet was typically high in sugar, processed food, and saturated fats showed significant improvement in mood, with reduced anxiety and depression, after eating a diet high in fruit, vegetables, fish, and olive oil. Nutrition has mostly been an afterthought for psychiatrists, though there is mounting evidence that nutrition is as important to psychiatry as it is to cardiology, endocrinology, and gastroenterology.

Knowing that good eating habits are important doesn't necessarily translate into real-life behavior. Many other factors influence decisions about what we eat. There was a scene in the popular film *When Harry Met Sally* when Harry and Sally are sitting in a deli, and Sally loudly fakes an orgasm to prove that men can't always tell when women fake it. The other diners listen and watch in amazement until an elderly lady tells the waiter, "I'll have what she's having."

Other people's behavior rubs off on us. Subconsciously we mimic gestures, adopt mannerisms, and even "catch" the moods of others. Consider what happens when you go to a restaurant with friends. If one of them orders a cheeseburger with fries, you may feel tempted to do the same. We like to think of ourselves as unique, but much of the time we're copycats. You might rationalize your choice by telling yourself, *If she's having it, it can't be that bad*. Or after a long day you might think, *I deserve a little indulgence, too*. Or you might worry about food FOMO (fear of missing out) and know that when your friend's order of the spaghetti carbonara arrives, you'll feel disappointed with your garden salad. Other people can also influence how much we eat. Eating with friends often leads to longer meals, and longer meals often means we eat more. Instead of relying on internal signals like a feeling of fullness, we turn into tumbleweeds, rolling with the wind and unwittingly controlled by external forces.

Restaurants have found that placing dishes at the top or bottom of the menu for a given course are up to twice as popular as those placed in the center of the list. I wish restaurant owners would leverage this information and nudge customers more toward healthier dishes. Elaborate descriptions also influence choice. Detailed menu labels with embellished narratives make our mouths water. "Slow cooked in our wood-burning oven with hand-picked rosemary and zucchini blossom" is more appealing than the straightforward "roasted with some tasty stuff." The euphemism "delicately fried" is my favorite attempt to downplay a less healthy choice. Descriptions that trigger happy memories are especially popular: "Grandma Ruth's secret-recipe home-baked potatoes" sounds much tastier than plain old "potatoes," especially if you're craving comfort food. Size also matters—the bigger the portion, the more we tend to eat. We tend to consume more potato chips from a jumbo bag and drink more soda from a Big Gulp.

Workday stress can also lead to overeating and unhealthy food choices. If we're not paying attention or being deliberate about what we eat, it's easy to board the devitalizing merry-go-round. Poor eating choices can make us more vulnerable to everyday stress, and everyday stress can lead us to make bad choices that increase our vulnerability. Feeling physically exhausted may be the reason we hear that doughnut calling our name. An argument with a loved one can bring on a feeding frenzy. When feeling busy and overwhelmed, it's all too easy to confuse our negative emotions with hunger pangs.

Feeling bitter literally makes us crave sweetness. Levels of the hunger hormone ghrelin have been found to increase following conflict with a partner. The more negative social interactions a person has daily, the more a person risks weight-related issues in the future. A study asked more than one hundred healthy young women to rate the quality of their daily encounters as negative (i.e., filled with criticism, disappointment, anger, and shame) or positive (i.e., filled with respect, increasing confidence, a sense of intimacy, and receiving helpful advice). The women who reported more negative interactions were more likely to have a larger waist circumference two years later.

These days I grill my patients about what and *when* they eat. The closer we eat to bedtime, the less well we sleep, which in turn may lead to cravings for unhealthy food the next day, not to mention fatigue and a bad mood. Timing matters during waking hours as well. Allowing too much time to pass between meals puts us at risk of becoming "hangry," that toxic combination of hunger and anger. If you have an empty stomach and encounter a negative situation, you might express "hanger" more quickly. There isn't a firewall between our mental and physical experiences. Feeling famished makes us irritable. Researchers analyzed parole hearings for inmates and found that judges were less likely to grant parole right before lunch than afterward. In fact, the percentage of favorable rulings dropped from 65 percent right after breakfast to nearly zero before lunch and then shot back up to 65 percent after the lunch break. Even experienced judges were swayed by an empty stomach. The decision to grant parole requires serious thought and consideration. When tapped out, they were more likely to deny parole, which requires less processing. Alas, the old saying, "Justice is what the judge ate for breakfast" rings true in the courtroom . . . and probably at the office and in the bedroom and in the line at the grocery store.

Physical sensations cross over into emotional ones and vice versa. When you're hungry or tired or under the weather, a molehill easily becomes a mountain. When you're upset or down, physical pain is even harder to bear, and shortcuts are more appealing. Pay attention to the signals your body sends you, because they can impact not only the quality of your day-to-day interactions and experiences but also your long-term mental health.

Relationships bear the brunt or reap the benefits of a healthy diet. Levels of irritability, responsiveness, patience, and interest in sex rise and fall, depending on what we eat and drink. A patient who switched to the Mediterranean diet didn't notice an immediate difference in his mood or energy level, but his partner did in just one week. "She says I'm back to my honeymoon self," he laughed.

I urge you to keep a diary in which you track your food consumption, mood, and energy levels, like the participants in the fruit and vegetables study. Then, look at the data and see if you can draw any conclusions.

And remember, as noted before, people are capable of multiple simultaneous changes in their mental and physical health. A study from UC Santa Barbara found that simultaneous changes often reinforce one another. Eating more fruits and vegetables, for example, will likely give you more energy, making it easier to exercise and get better sleep. "Helping people make progress in many ways . . . creates an upward spiral where one success supports the next," said Michael Mrazek, the lead researcher.

Even when weight isn't an issue, it is rare to meet a patient who never thinks about the scale and whether it's tipping up or down. Caring about how they look naked, how their jeans fit, how they look in the mirror are not superficial concerns, as vanity also affects how they feel. Because some of the best medicines we have for mental illnesses can also cause weight gain, it's no wonder that patients are not always compliant. I work with them to treat their illness while also honoring how they feel in their body. Being more deliberate about diet can positively impact one's everyday existence.

PART FIVE

Beyond You

Chapter 20

CONTRIBUTE VALUE

The waiting rooms of psychiatrists are filled with patients who feel aimless and empty. In an effort to lift their spirits, they often prioritize self-immersion over connection. Instead of living full, embodied, and deliberate lives, they look for shortcuts while pursuing the ultimate goal: a carefree life.

Now, a carefree life might sound appealing, but it's often our caring for others—and about causes—that fortifies us. When we care, we are engaged. We participate. We dig in. We persist. Investing time and expending energy beyond the immediate self can be challenging and stressful. It is also gratifying and sustaining. When you extend, expand, and stretch yourself, you are more likely to create enduring psychological resources.

Margot

One spring a patient named Margot came to my office feeling despondent. She had recently attended a seminar on the importance of self-care, which was titled, "Make This Year All about *You*." The two-hour workshop emphasized how prioritizing oneself was a must to achieve happiness. Margot was instructed to put herself at the top of her to-do list and begin each morning by looking in the mirror and asking, "What do I need today?" She was told to make regular "dates" with herself and to treat herself with a slice of her favorite cake or a manicure. At the end of the seminar, the attendees signed a contract pledging to give more love, kindness, and attention to themselves.

When Margot returned home, she withdrew from her book club so she could read the recommended self-help books. (Plus, she told me, because the club didn't always choose books she liked, she felt further justified in her decision.) The signed pledge gave Margot the license to decline invitations that weren't convenient or to her liking. She decided not to attend a friend's birthday dinner because it wasn't being held at a vegan restaurant. When her sister came to town for a visit, Margot barely made time to see her.

Margot focused on herself at the expense of other relationships and did see some positives. She was getting lots of sleep, eating a healthy diet, reading a self-help book a week, meeting with a life coach every two weeks, meditating thirty minutes a day, and getting plenty of exercise. For her vacation she canceled a visit to see her grandmother and opted instead for a silent retreat. Yet in spite of her efforts to give back to herself, Margot said that her efforts hadn't provided the gain in happiness she had hoped for. If anything, she told me in almost a whisper, she felt worse.

I let Margot know that she wasn't alone in feeling let down. Putting yourself on a pedestal can actually erode well-being because it greenlights self-focus and cuts people off from others. I told her about an experiment in which volunteers were asked to choose one of three acts to perform each week for a month: to show kindness to other individuals, to humanity, or to themselves. The groups that performed acts of kindness toward others or toward humanity experienced a greater boost than those who focused on themselves. A massage is relaxing and enjoyable in the moment, but the positive feeling fades quickly. When acts of kindness are other-oriented, not self-oriented, people feel better for longer. The study concluded that when "people do nice things for others, they may feel greater joy, contentment, and love, which in turn promote greater overall well-being and improve social relationships." In short, a cascade of uplifts follows other-oriented actions—and they linger.

People also tend to feel better when they buy a gift for someone else than they do when they buy it for themselves. Plus, the happiness derived from giving things does not wear off in the same way as purchasing something for oneself does. *Having* can get boring but the "warm glow" of *giving* sustained itself over the course of the study. Altruistic behavior can even dial down anxiety. People who are socially anxious were able to override feelings of insecurity and feel more confident after actively lending a helping hand, such as by mowing a neighbor's lawn or doing a favor for a roommate. Self-care might be all the rage, but it's important not to forget "other-care" as a source of vitality and resilience.

"What about the gratitude exercise I do every night?" Margot asked. "Shouldn't that be making me happier?"

I asked her to elaborate. The seminar had told the attendees to list the things that made them grateful before going to sleep. Margot hadn't missed a day, and I asked what had made her list the night before. Margot said she was grateful for a sweater her mother had given her: "I got so many compliments on it." She was also grateful to a coworker for filling her in on the contents of a meeting she had missed. "It made me feel in the loop."

Margot was dwelling on how these acts made *her* feel better, but research shows that the power of gratitude lies in expressing it toward others. I explained that there are two types of gratitude: "other-praising," which recognizes someone else and strengthens social bonds, or "self-benefit," which focuses on what the recipient has gained. Being grateful *to* her mother for being so thoughtful and *to* her coworker for being so helpful are examples of other-praising gratitude. Feeling "in the loop" and complimented is satisfying, but we lose the magic of gratitude when we make it all about us.

People who express gratitude toward others have stronger and more loving relationships. So, if your partner sends you flowers, you can relish how great the gesture made you feel, or you can channel your gratitude toward your partner by actively saying or doing something that acknowledges how awesome *she* is. It is the difference between saying, "Thank you for the flowers; they cheered me up," versus "Thank you for the flowers. You cheered me up." Think of showing gratitude as an expression but also as an action—as a verb that works best when it is embodied, spoken aloud, and when it connects you to someone else.

Margot wanted to be happy, but her quest had led her down an unfulfilling path of self-immersion. She had been seeking growth and connection. What she found was pseudogrowth and isolation. I explained to Margot that the more people value their personal happiness, the lonelier they feel on a daily basis.

"I guess I'm exhibit A," she acknowledged.

Neuroscientist John Cacioppo has explored how loneliness leads to self-centeredness, developing a theory that has an evolutionary basis: for our ancestors, the experience of loneliness functioned as a warning system to take care of one's immediate welfare and interests. "In modern society becoming more self-centered protects lonely people in the short term," says Cacioppo, but the more self-immersed they become, the harder it is to reconnect socially.

Loneliness and self-centeredness create a feedback loop wherein one bolsters the other. This can lead to social isolation, which in turn can damage mental and physical health. When people are actively generous with others, they typically feel more competent in their ability to add value, to enact change, and to feel that they belong. Of course, it is important to take care of yourself, but don't let self-care become an excuse for self-immersion.

Hold the Door

Paying too much attention to what's going on in our own heads can lead us to ignore stress-buffering experiences, connections, and opportunities. Jumping in and doing things for others is one of the most effective antidotes for everyday stress. Assisting a neighbor with groceries, giving a compliment, bringing a sick friend a bowl of soup, visiting grandparents, and volunteering at a library or shelter are all actions that can help us feel less rattled by the barrage of pebbles that get caught in our shoes.

Being kind and generous offers many benefits, including better relationships, greater confidence, and a stronger immune system, as well as improved cardiovascular health and a decrease in physical pain. Acts of kindness can even lessen symptoms of depression. And they're good for business. Having an other-orientation may make us better at our jobs . . . doctors included. Researchers at Stanford University found that having an empathetic doctor can influence a patient's outcome. Volunteers were pricked on their forearm with histamine, which caused their skin to become red and itchy. The volunteers were then briefly examined by either a friendly doctor, who offered a few words of encouragement, or a reserved doctor, who assessed the reaction but did not offer any reassurance. The volunteers who experienced a warm interaction reported less itchiness and their symptoms resolved more rapidly than those who were seen by the aloof physician. We all know that competence is important but caring is too.

If the quality of interactions didn't matter, then both sets of patients would have been expected to report similar experiences. A doctor who calls you by name, who looks you in the eye, who smiles and chats, and who expresses genuine concern for your well-being can help you heal. A doctor who rolls his eyes at your complaints, or brusquely tells you "you're fine" when you're not feeling so, may have been at the top of his class but also may actually be bad for your health.

Have a Time Feast

If daily demands make it difficult to take care of yourself and help others, you have company. Nearly 50 percent of Americans say they don't have enough time, an epidemic that has been described as "time famine." This pervasive feeling of having too much to do but not having enough time to do it chips away at our well-being.

It may be counterintuitive, but research shows the best way to feel as if you have more time—"time affluence"—is to give it away. Cassie Mogilner, who studies decision-making, led four experiments that showed people's subjective sense of having time increases when it is spent on others. Being generous with your time (for example, by lending a hand to a friend, taking a neighbor's dog for a walk, or through volunteering) boosts a person's belief in his capacity to add value and make a difference, which shapes his perception of time. Mogilner and her co-researchers concluded, "We identify a specific choice that individuals can make to lessen their experienced time pressure: Be effective by helping others."

Forget the Mask Metaphor

"In the unlikely event of a loss of cabin pressure, oxygen masks will drop down. Secure your own mask first before helping others."

We all know the drill. And if you ever forget it, a flight attendant will remind you at the beginning of every plane trip. The advice makes sense. If you're trying to help someone but you run out of oxygen, both of you could wind up losing consciousness.

In the field of therapy, the oxygen mask has become a metaphor for prioritizing the self at all times and above all else. The message is loud and clear: Focus on your own needs. Look out for number one. Everything else can wait. Or as Gwyneth Paltrow put it in 2020 in *Town &*

Country magazine: "I think what this wellness movement is really about is listening to yourself, tuning into what interests you, and trying things. Find what makes you feel better and go from there."

This message lands easily on receptive ears. People tend to veer in the direction of self-interest when they are stressed out. Run-of-the-mill, everyday worry is enough to increase an egocentric point of view. In an admittedly odd study, participants were asked to think about a moment when they felt anxious. Next, they were shown the photograph below.

Participants were then asked, "On which side of the table is the book?" The people who were made to feel anxious were more likely to say, "The book is on the right." They answered using their own perspective instead of adopting the perspective of the person in the photo.

When feeling stressed, it is harder to put yourself into another person's shoes. I recently saw a patient who told me about a tense interaction she'd had with a man whom she'd been dating for a few weeks. As dinner was winding down, he took her hand. She recognized it was intended to be a romantic gesture, and yet it bothered her. He had finished eating, but she hadn't. To her, wanting to hold her hand felt self-absorbed, not affectionate. Wanting a few more bites of her hamburger, she had pulled her hand away, and an argument followed. She told him he was selfish; he told her she was overreacting. They hadn't spoken since. I couldn't help but think of all the moments in which we remain stuck in our own minds and oblivious to those around us. My patient's experience is also a reminder that communicating your feelings clearly can rescue these moments. If she had said, "You're sweet but I really want to finish this burger," they might have both laughed.

In the parable of the Good Samaritan, a traveler was robbed, beaten, and left to die on the side of the road. Two religious men—a priest and a Levite—pass by, crossing to the other side to avoid him, while a lowly Samaritan stops and helps. Jesus tells the story to illustrate what it means to be a good neighbor. Were the men who ignored the traveler coldhearted or just in a rush? A classic study conducted by two Princeton psychologists highlights how time pressure can turn us inward, blinding us to the needs of others. Seminary students were asked to prepare a three-to-five-minute talk on either what it means to be a minister or on the topic of the Good Samaritan. They were then instructed to go to a building across campus to finish the project. Some of the students were told to take their time; others were told they were late and needed to hurry. On the way, each student encountered a "victim" slumped in a doorway with his head down, eyes closed, not moving. As the student went by him, the victim coughed twice and groaned, keeping his head down. One might expect that the students assigned to give a talk on the Good Samaritan were more likely to stop, but that wasn't the case. In fact, the researchers noted that some of them literally stepped over the victim. Their reaction wasn't necessarily a reflection of their callousness but rather of their sense of urgency, which in this situation eclipsed empathy. As the researchers observed: "One can imagine the priest and Levite, prominent public figures, hurrying along with little black books full of meetings and appointments, glancing furtively at their sundials."

One reason spirituality is thought to enhance mental health is that it reduces self-centeredness and creates a sense of belonging to a larger whole, regardless of the particular faith. Despite differences in rituals and beliefs among religious traditions, living life according to a value system that emphasizes an other-orientation can often boost feelings of well-being.

Contribute a Verse

As tempting as it is to believe that "I-talk" is a litmus test for narcissism, the evidence isn't there. People who frequently use the pronoun "I" and other first-person pronouns are more easily upset and prone to woe-is-me thinking. But they are not necessarily insufferable narcissists. The problem with I-talk is that it places you in the psychological spotlight, opening the door for self-immersion. Other people's negativity feels personal and aimed directly at you when you're in "I-mode." Another issue with I-talk is that it can be off-putting to others or—to be frank—boring. It is hard to meaningfully connect with someone who seems preoccupied with his own experience.

I-talkers are often the last to know they are I-talkers. Ask family and friends to point out when you are I-talking if you want to be more aware of it. Also, if you're telling a story or typing an email, pay attention to how much you use the words *I* and *me*. If you catch yourself in the act, frame the conversation differently. Instead of saying, "I can't believe how nasty the salesperson was to me," opt for, "The salesperson was rude." Or perhaps try to see the experience from the other person's perspective: "Maybe the salesperson was having a really bad day." Deliberately engaging in conversation and activities that help you to go beyond yourself and connect with others is more vitalizing as well as more interesting.

Walt Whitman's poem "O Me! O Life!" might be the perfect antidote for I-talkers. It opens with an exasperated tone that calls out "the plodding and sordid crowds I see around me," Feeling "empty" and "useless," the poet poses the existential questions—and I paraphrase—"What's the point of it all? Why am I wasting my time?" He comes to the conclusion that the point is to contribute. Whitman expresses his "answer" as follows (emphasis mine):

> "That you are here—that life exists and identity,
> That the powerful play goes on, **and you may contribute a verse**."

What verse will you contribute to this ongoing powerful play? Parker Palmer points out: "No one ever died saying, 'I am sure glad for the self-centered, self-serving, and self-protective life I've lived.'" That's harsh, but it makes an important point. To be clear, I am not promoting self-neglect or martyrdom. Nor am I recommending a life that rivals that of a doormat. It is important to take care of yourself. In the event of an emergency at thirty thousand feet, putting on your oxygen mask first could make the difference between life and death. By all means, secure yours before helping someone else. Still, when it comes to everyday life, this metaphor usually doesn't fly. Plus, it doesn't have to be either/or; it can be both/and. Too much self-focus can become an excuse to shut ourselves away from the rest of the world. There is nothing wrong with doing good things for yourself, but taken to the extreme, it can turn into a justification for self-absorption.

Chapter 21

MAKE YOURSELF USEFUL

When my sister and I were children, we never dared to use the "b word" around my mother. To admit we were "bored" was considered blasphemy in our house. My mother had a go-to command whenever she suspected her children were not making good use of their time. If we complained that there was nothing to do on a hot summer afternoon, she would inevitably say, "Go and make yourself useful."

Making ourselves useful meant to weed the garden, sweep the gravel, clean the garage, pick rocks out of the drain, wash the car or, to do something, *anything*, that would shake us out of our self-absorption.

Weeding the garden was especially frustrating and tedious. Those relentless creepers were stubborn and rough. They hurt my fingers, while layers of soil would take up residence under my nails, requiring days of brushing to remove. Still, there was also something gratifying about the process. As I filled the bucket up with the tendrils of these earthbound jellyfish, I stopped ruminating about whatever was bugging me. The ticker tape of self-focused thoughts took a break as I dug my fingers into the earth. Over time I got better and better at weeding. Instead of just pulling out the heads, I learned that with the right amount of tugging, I could get the whole thing, including the root, in one swoop. Sure, the weeds would grow back, but making a tiny difference in that moment gave me a lift.

My mother was keenly aware of the extraordinary transformative power of making yourself useful. Human beings may veer into self-interest, but we also have a strong inclination to contribute. Young children will go out of their way to lend a hand, even to someone they don't know. During one study, when an experimenter dropped a pen or struggled to open a cabinet, toddlers immediately tried to remedy the situation. Related research showed that two-year-old children find helping others as enjoyable as helping themselves. Adding value fulfills a fundamental need for growth, fulfillment, and connection. Above all, it helps the person feel *valuable*.

Experience Corps, a program that pairs older adults with young students to provide academic support in urban public schools, is one striking example of how this principle has been applied successfully on a wide scale. Regularly meeting with elderly tutors boosted elementary students' reading skills and test scores. But the children weren't the only ones who benefited from the mentoring. After a year with Experience Corps, 84 percent of Experience Corps volunteers reported that their circle of friends—a key measure of social well-being, particularly for aging adults—increased as a result of their involvement in the program. About two thirds of the least physically active older adults reported that they became significantly more energetic and felt more engaged in social and community events. Most significantly 86 percent of Experience Corps volunteers said their lives improved because of their involvement with the program. A renewed sense of purpose in life and connection lies at the heart of these improvements.

Make It Not about You

One adage claims that the three essentials of a good life include "something to do, someone to love, and something to hope for." I would add that you also need something to do *that matters beyond you*.

John had wanted to quit smoking for years. He had tried everything from hypnosis to prescription medications to nicotine patches. In the short term, the interventions worked, but after a month or so he was back to a pack a day. Another therapist encouraged him to dive into his past. John's father smoked a pipe, and perhaps that connection would offer insight about why John smoked and help him to stop. This approach didn't work either.

So much of our well-being and mental health is thought to rest on these three areas of focus:

1. On the past (e.g., what has happened), rather than on the future
2. On the individual, rather than on how a person exists in relation to others
3. On thinking, rather than on action

Yet none of these principles was helping John quit smoking, and he remained frustrated. In sessions with me, he would cite the reasons why he wanted to quit: Vanity. (He hated the crinkles developing around his mouth.) Health. (He was prone to upper-respiratory tract infections.) Odor. (His apartment and clothes reeked of smoke.) Still, recognizing all the personal benefits he would reap wasn't enough to break the habit. What ultimately made John quit for good was his four-year-old nephew.

After a family meal, John would typically excuse himself from the table. "I'm going to go outside to get some fresh air," was code for "going to have a smoke." One evening little Lucas stood up and said, "Uncle John, I'm coming, too!" Then the boy grabbed a breadstick and mimicked smoking. That was it for John. Wanting to set a good example for his nephew reignited his motivation. He stopped smoking that day permanently.

Psychologists call this "a self-transcendent motive." Recognizing how your actions and experience impact someone else can shift behavior. When asked about why they wanted to go to college, students from low-income families who agreed with statements like, "I want to learn things that will help me make a positive impact on the world" and "I want to become an educated citizen that can contribute to society" were less likely to drop out than those who expressed self-oriented motives like, "I want to learn more about my interests." Students that were asked to consider how learning relates to making a positive contribution were more likely to complete "boring" math problems than watch viral videos or play Tetris. This "diligence task" was designed to mirror real-world choices students face every day when doing their homework. Having a motive that transcended their personal concerns helped them persevere.

To illustrate this point I like to tell patients that when St. Paul's Cathedral was under construction, the eminent architect Sir Christopher Wren asked three laborers at the site who were performing identical tasks what they were working on. One responded, "I'm laying bricks." The second said, "I'm building a wall." The third said, "Sir, I'm building a cathedral." Seeing his daily work as part of a larger effort to build a place of worship for the community imbued the third man's task with meaning. The daily grind can turn us all into bricklayers. Deliberately connecting our actions to something bigger keeps us vital and reminds us to preserve the image of the metaphorical cathedral in our mind's eye.

When the going gets tough, the motivation to continue often weakens. Turning an activity into a mission can provide a second wind. As John, the former smoker commented, "It's a whole lot easier to answer the question, 'Why am I doing this?' when the reason is about more than me."

People often become defensive when they are reminded of their unhealthy behaviors. They would rather plug their ears and loudly sing a song than listen to advice. One of the reasons they close themselves off from hearing important messages about health is that these messages feel to them like an indictment of their choices. Shifting focus away from the self tends to make people more open to the idea of making a change. A friend who is an internist was lamenting how patients became defensive whenever she spoke to them about stopping smoking, going on a diet, or exercising more. "It's like 'talk to the hand.' They just don't want to hear it," she explained.

She has since found that leading patients to share what is meaningful to them or what gives them a sense of purpose makes them more receptive to a conversation about effecting change. "Sometimes I'll just say, 'Tell me about your grandchild' or something I know that they care a lot about. If, a few minutes later, I say, 'Let's talk about your cholesterol level,' they're much more open to the idea of cutting down on eating burgers and fries."

A study conducted by Adam Grant of Wharton and his colleagues found that the best way to encourage hand hygiene among health care professionals isn't to scare them about the personal consequences of not washing their hands, but instead to remind them of the potential consequences to others if they don't. Though doctors and nurses are well aware that handwashing is vital to prevent spreading germs, many become busy (like the men in the Good Samaritan story) and forget to make it a priority. So the researchers posted two different reminder signs at handwashing stations. One read, HANDWASHING PREVENTS YOU FROM CATCHING DISEASE. The other read, HANDWASHING PREVENTS PATIENTS FROM CATCHING DISEASE.

Health care workers used 45 percent more soap at the station with the sign warning about the danger to others than at the station that appealed to maintaining their own health. Tapping into what is meaningful for people—certainly caring for patients in a hospital—seems to be a stronger motivator than the threat of personal harm and self-interest. The study results concluded, "Safety behavior is not necessarily 'all about me.'"

Give and Receive

Parents frequently ask me about how to engage a disengaged student. Full disclosure: it's a question I'd like an answer to myself. Getting my children to do homework is no easy feat. Punishment is one strategy, but how many times can you take away their phones? Rewards are another option, but research shows that this is not necessarily a good idea. Expecting something in return for putting in required effort can actually undermine motivation. If you incentivize your children to get good grades with cash, it's unlikely they will ever find any internal motivation to want to do well on their own.

My go-to strategy has always been to explain why working hard is important and then to offer advice about establishing solid study habits. Having worked my own tail off in school, I consider myself to be a treasure trove of invaluable information about how to buckle down and get it done. For years I thought my words of wisdom would light the academic fire within my children. Perhaps seeing their eyes glaze over while I banged on about doing "the maximum, not the minimum" should have alerted me that the message wasn't sinking in.

Contrary to the assumptions of well-meaning parents and teachers everywhere, explaining to children why and how they should study doesn't make much of an impact. Most children are fully aware of the value of an education and optimal study habits. They don't need more information; they need more motivation. Research offers a counterintuitive solution. Instead of giving students expert advice about how to do well in school, ask them to provide this advice to other students. Middle-school pupils who shared their thoughts about the importance of doing vocabulary homework with fourth graders became more motivated to study vocabulary themselves.

Giving advice to others boosts confidence. Instead of being a struggling student in need of guidance, the student becomes a person with valuable experience capable of providing assistance. Plus, people like to be consistent. When advocating for an idea, we take ownership of it. In the process of explaining how important something is, we persuade ourselves of it, too.

The motivational power of giving advice isn't just applicable to students. People trying to lose weight, control their temper, save money, and find a new job became more committed after giving advice to others facing the same effort. Helping others tapped into their own motivation to be successful. Most adults are fully aware of the steps required to achieve their goals. Traversing the gulf that separates knowledge from action is the challenge. As these studies suggest, flipping strugglers from receivers into givers provides a bridge and also the fuel to enact what you care about.

So next time you encounter a person or a child who is having trouble reaching a goal, save your breath. Instead of offering your own words of wisdom, ask them what they would say to another person in a similar predicament. In giving, they will receive.

Chapter 22

DELIBERATE VITALITY

As a child I couldn't understand why the nurse in my pediatrician's office was so mean to me. Did she have any idea how much those shots hurt? She would stab me and then cheerily exclaim, "All done!"

But it wasn't. My arm would ache for hours afterward. Apparently one time I decided to fight back. According to family lore, I bit her, although I swear that I don't recall doing this. But if I did, in my defense, I truly believed that she was a heartless woman who relished jabbing needles into the spindly upper arms of defenseless sweet children.

It is not always obvious that someone means well. I now understand that the nurse had my best interests at heart. Still, people's intentions are often ambiguous. In the absence of clear communication, it's easy to misinterpret the actions of others. I wish I could say that this tendency ended for me on becoming an adult, but recently I emailed a friend with a slightly awkward question on a Monday. My friend didn't respond the next day. Or the day after. I became convinced that she was furious with me. I thought about writing a follow-up to apologize but feared I'd only make the situation worse. When she finally did write back that Friday, she explained that she'd been swamped at work, and then happily answered my question and signed the note with "xo."

One of the many questions psychiatrists are trained to ask new patients as a way to assess paranoid ideation is whether they ever feel that people are out to get them. Paranoid ideation is very different from the feeling that someone hates you because they didn't return an email. A person experiencing paranoid ideation might interpret a seemingly benign situation as confirmation that they are being targeted, harassed, or treated unfairly. They might think someone walking in the same direction is following them, or that a random person talking into his phone is saying negative things about them. Bird poop landing on their head could be proof of a conspiracy. Paranoid ideation may suggest an illness like schizophrenia or schizoaffective disorder.

I have several patients who don't have paranoid ideation but have what I call "pessimistic ideation," a tendency to see the world through a dark lens. Any negative experience is seen as further evidence that the world is a hostile place full of selfish people. From their perspective, every piece of data fits into this pattern. Often these negative perceptions intensify when they are feeling exhausted or overwhelmed. Daily stress reduces the willingness to give others the benefit of the doubt. Suddenly an ambiguous interaction or small mistake feels like a major betrayal. If the barista doesn't put enough milk in their latte, she is labeled as incompetent. If a spouse forgets to pick up the laundry, he is a narcissist. The more stressed we are, the less patience they have. Eye rolls, cold shoulders, and exasperated sighs are the lingua franca of irritated people everywhere.

Assuming negative intent makes a bad situation worse. For one study, three groups were given unpleasant electric shocks administered by a partner. The first group was told that the shock was delivered by accident and outside their partner's awareness. The second group was told that they were being shocked intentionally, but for no particular reason. The third group was told that they were being shocked because their partner was trying to help them win lottery tickets. Those in this third group reported significantly less pain than those in the other two groups. Simply knowing that someone had good intentions and was trying to help them lessened the pain. (This study made me wonder if my pediatric nurse had taken the time to explain why she was giving me a shot, it wouldn't have hurt as much and I wouldn't have bitten her. A win-win.)

Be Deliberate and Follow Through

Take the time to communicate your good intentions. Nobody can read your mind, and no issue ever got resolved by a conversation you had in your head. As Maya Angelou said, "You may have a heart of gold but so does a hard-boiled egg." The actions we take to treat people with kindness, to express interest, to give someone our full attention, to listen, and to learn shape the valence of our interactions and the quality of our relationships. When you have an impulse to show someone that you care, act on it. Write that card . . . express appreciation . . . show respect . . . say the words you're thinking. Feeling loved and respected in everyday life make for better days and contribute to greater well-being. Go out of your way to cultivate moments of felt love for others. Be deliberate about how you relate to others.

And be deliberate about what you hope to achieve for yourself. Hold yourself accountable by frequently measuring and monitoring progress. If you want to make a change, engage in behaviors and accept challenges that help you meet this goal. Desiring change or making a plan in your head is not enough. Setting goals that are specific, measurable, action-oriented, realistic, and time limited can better enable you to enact them.

For instance, if you would like to be more organized, don't just make a plan to clean up your desk, implement it. If you want to become more punctual, show up ten minutes early for your next appointment. If you intend to be more open-minded, engage in a conversation about a topic with someone who doesn't agree with you and listen. If you would like to wake up twenty minutes earlier tomorrow morning but worry you will press the Snooze button on your alarm clock, put the clock across the room. If you would like to feel less critical of your partner, ban the word *should* from your vocabulary for an entire weekend. Wanting things to be different without taking action will leave you wanting.

Decide to decide about what matters to you and make the choice to embody these values on a daily basis. Actively engage in behaviors that make you feel strong—be deliberate about what you eat, about exercise, and about sleep. Actively pursue meaningful uplifts. Go out of your way to treat others with respect and kindness. Be intentional with your attention. Override the urge to do what's easy but empty. Embrace opportunities to explore and expand. Overrule the impulse to avoid challenges and retreat. Savor meaningful moments. Share them with others. Seek and generate moments of felt love. Be actively generous. Resist the temptation to fixate on what's wrong or who has wronged you. It sounds like a lot . . . but the more your actions align with your values, the more vitality you will have.

Vitality is not an outcome. It is a process. Cultivate it by making choices that deliberately reflect what matters to you and that build competence and connect you to others. Embody what you care about. Set realistic goals for yourself—and not just "what" goals with a specific outcome in mind but with a "why" behind them as well. The "what" will focus your energy. The "why" will replenish it, because it taps into what you care about.

Reimagine Resilience

As I was finishing this book, COVID-19 continued to upend our lives. A patient with a history of anxiety told me that although her anxiety was under control, she still felt off kilter: "What I'm feeling is hard to put into words. It's not my regular catastrophic worry—it's more of a dull, nagging, uneasiness that occurs when something you love goes missing."

The unfamiliar sensation many were experiencing was grief—grief for what was missing and also anticipatory grief, an ongoing dread that more loss was yet to come. Uncertainty amplified the unspooling heartache. Ceaselessly checking for COVID-19 news fueled fear, as if checking on the number of people hospitalized would make a difference. Overloading on news is always a risk. Sensationalized media coverage takes a toll on mental health, amplifying existing fear and fueling fear about the future. A study conducted after the Boston Marathon bombing in 2013 found that people who experienced the most distress were not those who had been in the vicinity when the bombs went off or who knew someone who had been injured. Rather, it was the people who were glued to the news. Consuming six hours or more of news in the week following the bombing predicted higher acute stress.

This relentless appetite for negative news stories reminded me of a study that explored satiety: participants who unknowingly ate soup from a self-refilling bowl consumed 73 percent more soup than those who ate from a single-serving bowl. Like the soup eaters, we mindlessly consume negative news, not realizing our brains are being force-fed. Taking more control of how and when we get our news is essential to staying mentally strong.

On top of truly life-threatening reports, the call for social distancing made this period especially hard. We are deeply social creatures, and making the most of face-to-face interactions has been a central theme of treatment within my practice. Presence, I would tell patients, is everything—until COVID-19. Overnight our lives moved online. Seeing friends, family, and colleagues through a screen provided a sense of connection but somehow was a constant reminder of what we had lost. It was as if we were all in Plato's cave, except we knew what we were missing—conversations without awkward silences, laughter without delay, heartfelt hugs, and physical presence.

With patients (and friends), I avoided asking presumptuous questions like, "Are you okay?" which implied fragility or suggested that I didn't think they had it in them to actually *be* okay. Instead, through telehealth sessions, I asked patients if they were experiencing any uplifts. I explained that amid the suffering and sadness, it was important to work overtime to find moments of positivity. Poet Ross Gay's *The Book of Delights* catalogues his moments of delight over the course of a year. Gay notes that the practice occasioned a kind of "delight radar." The more he studied delight, the more he found it. It is not that sorrow or fear disappeared from his life or were diminished by the search for joy. What he learned was how suffering and grace can live side by side. And if we don't deliberately seek out grace, all we see is darkness. Intentionally bringing the unseen and the underappreciated into view is an act of generosity that can pull you out of yourself.

During the pandemic, every time we opened our computers, checked our phones, or turned on the TV, we saw suffering and loss. Yet, we also bore witness to its opposite—generosity, goodness, and compassion. A son sat outside his father's nursing home window to make sure his dad saw him every day. A man held up a sign that read, THANK YOU ALL IN EMERGENCY FOR SAVING MY WIFE'S LIFE. I LOVE YOU ALL. Health care workers were applauded from balconies and rooftops. Neighbors took care of neighbors.

To help protect patients from feeling completely overwhelmed by the endless barrage of bad news, I advised them to deliberately cultivate or take notice of at least two ordinary moments each day that were uplifting. I suggested they jot down a sentence about it or take a photo to ensure the experience registered, so that it wasn't just noticed in passing and then forgotten. Sharing the moment with another person was another way to guarantee that those ordinary moments left an imprint. As Gabriel García Márquez said, "What matters in life is not what happens to you but what you remember."

Some patients felt a sense of guilt about allotting time for uplifts, as if experiencing joy was somehow a betrayal of grief. Other patients who were doing fine confided they were embarrassed that they weren't more distressed or overwhelmed.

One patient with severe anxiety had a surprising reaction to the stay-at-home order. "I'm actually doing pretty well," she told me. It helped that she was applying all the strategies she had learned over the years to cope with the anxiety—to get enough sleep, limit news and social media, eat healthy, go for walks once a day. "It's like I have been preparing for this all my life," she said. "I'm actually coaching my typically nonanxious friends, who are insanely anxious right now, about how to handle all the uncertainty. I tell them, 'Welcome to my world.'"

Look Forward

Twenty years of practicing psychiatry has taught me that people CAN and DO change and also that excavating the past—rehashing problems from the previous week or decades ago—is rarely a game changer. Reflecting on your own life, it probably seems obvious that you have changed a great deal. I suspect that you are not the person you were when you were twenty-one. (I am not just talking about your fashion choices.) I doubt you're exactly the same person you were two years ago, and possibly even six months ago. While in college, I read Edith Wharton's *The House of Mirth*. Back then, I had underlined what I thought were "the important parts" (just as I do now), scribbling notes in the margins, dog-earing what resonated. But when I recently reread my old college copy of the novel, I discovered new important parts. What spoke to me now was not what had spoken to me then.

You can recognize that your present self has evolved, but it's a lot harder to predict how you will change in the future. Still, you will. And if you are deliberate about the changes you want to bring about, you have a better chance of achieving them. This is why I often ask my patients, "Moving forward, how would you like to feel next week when you come in? Can you choose one small thing that can help push you toward that?" We cannot control everything, but there's plenty that we can. Do not bend with the wind unless you approve of the direction it is blowing.

So much of our lives take place on autopilot. We mindlessly go about our days, squandering our attention and time. We overfocus on self-thoughts and wind up feeling disconnected and disengaged from what truly matters. Often, by being more deliberate, we can correct for this. I urge you to remember that going about your days *is* going about your life. Vitality is not the by-product of a passive existence. Contrary to what lifestyle gurus tell us, well-being is not up to the individual nor does it occur in isolation or require prolonged self-focus or a silent retreat. What well-being *does* require is daily effort.

Some wellness gurus have led us to believe that looking in the mirror is the best way to grow and cultivate vitality. Practicing psychiatry has shown me that we flourish more when we turn away from the mirror and look out the window. Even better, I hope you'll throw open the door and venture out safely into the world. Choose vitality by challenging yourself, connecting with others, and contributing in all the ways that you can.

I wish you all the best.

ACKNOWLEDGMENTS

A web of extraordinary people made *Everyday Vitality* possible and I am immensely grateful to all.

Nell Scovell, you are an alchemist. In your hands, an idea becomes prose. You added precision, eloquence, and heart to each sentence. You also had an uncanny instinct for what to remove. Thank you for challenging me, inspiring me, and making me laugh. Your masterly mix of support, questioning, ideas, enthusiasm, and encouragement enriched every page.

Pilar Queen, my agent and friend, thank you for having a vision for this project when I did not see one. Your superhuman patience and unending support have been a wellspring of confidence and inspiration. Rick Kot, my editor and book whisperer, I am forever grateful for your insight, guidance, flexibility, and humor. As you know, I can be rather long-winded and you were the perfect windbreaker. Your idea to focus on everyday life helped me to expand my thinking while narrowing my focus. I am thankful to copy editor Francesca Drago and the dream team at Penguin: Lydia Hirt, Brooke Halsted, Shelby Meizlik, Carlynn Chironna, Fabiana Van Arsdell, and Cassandra Garruzzo. Special thanks to the indefatigable Camille LeBlanc for shepherding me through the publishing process and Lynn Buckley for visualizing vitality. Amy McWalters and the rest of the team in the UK, what a pleasure it has been to work with you. Bevin Lee, Sophie O'Rourke, and Heather Catania, thank you for helping me tell this story. Ben Fingeret, thank you for always going above and beyond.

Chapter 13 is about the importance of "Tailwinds"—the encouragement that allows us to grow and be more vital. This book would not exist if it weren't for the encouragement of Tory Burch. Once, during a walk in Central Park, she said, "Sam, all those studies that you email me about behavior and psychology? I think there's an audience for these conversations and ideas. You're a doctor with the latest research at your fingertips—why don't you figure out a way to share it with more people?"

"That would be great," I said. "But I just don't know where to start." Ten days later, Tory called. "It's time to act," she said. "Can you come into my office tomorrow?" I started to hedge, but Tory cut me off. "It's settled. Two-twenty tomorrow." The next day, I met with her team. They gave me an assignment to write a piece for the Tory blog about Mother's Day. My own blog, positiveprescription.com, was up two months later with the mission to bring science-backed information about positive mental health. I owe so much to Tory, who is not just my tailwind, but also a tailwind for so many.

I am blessed to have many tailwinds in my life. Jessica Yellin, where to begin? Thank you for decades of friendship. You are the ultimate tailwind. Murali Doraiswamy, thank you for the early read and invaluable feedback. Dilip Jeste, your support and wisdom have made me so much wiser. Angela Duckworth, thank you for so generously sharing the challenges you faced writing your book. Bearing witness to your grit inspired me to find my own. Arianna Huffington, thank you for your bountiful support and encouragement. Adam Grant, you are a giver *par excellence.* Jared Cohen, thank you for challenging, nudging, and cheering me on. James Pawelski, MAPP transformed how I think about well-being, especially the role of the humanities. Thank you for shining a light on what makes life worth living. Martin Seligman, you transformed how I think about patients and mental health.

This book benefited immeasurably from great conversations. Thank you Fernanda Niven, Hamilton South, Deeda Blair, Ruzwana Bashir, Eric Schmidt, Bob Colacello, Caroline Weber, Jessica Seinfeld, Wendi Murdoch, Billy Norwich, Manuel Bellod, Julie Frist, Barbara Tisch, Marjorie Gubelmann, Derek Blasberg, Nick Brown, Dambisa Moyo, George Makari, Lucy Danziger, and my extraordinary sister, Serena Boardman. "Ah, good conversation—there's nothing like it."

A heartfelt thanks to Mel Saldana for your thoughtful input and advice.

I am grateful to my parents, Pauline and Dixon, for abiding encouragement and support. Gaby, Charlie, Baker, and Vivian, you are a gusher of vitality. I have learned so much from all of you. This book is my effort to bring more vitality into people's lives. Thank you for bringing so much into mine. Aby, catching a glimpse of the world through your eyes is revitalizing. Thank you for the love every step of the way.

NOTES

INTRODUCTION

x **The answer is vitality:** Richard M. Ryan and Christina Frederick, "On Energy, Personality, and Health: Subjective Vitality as a Dynamic Reflection of Well-Being," *Journal of Personality* 65, no. 3 (September 1997): 529–65, https://doi.org/10.1111/j.1467-6494.1997.tb00326.x.

x **Andrew Solomon observed:** Andrew Solomon, *The Noonday Demon: An Atlas of Depression* (New York: Scribner, 2001), 443.

xiii **In his book *Flourish*:** Martin E. P. Seligman, *Flourish: A Visionary New Understanding of Happiness and Well-Being* (New York: Free Press, 2011), 54.

xiv **The levels of happiness felt:** Dilip V. Jeste et al., "Why We Need Positive Psychiatry for Schizophrenia and Other Psychotic Disorders," *Schizophrenia Bull*etin 43, no. 2 (March 2017): 227–29, https://doi.org/10.1093/schbul/sbw184.

xiv **"the most stressed workers":** Mental Health America, "The Mental Health of Healthcare Workers in Covid-19" (survey), accessed January 25, 2021, https://mhanational.org/mental-health-healthcare-workers-covid-19.

xv **Everyone deserves to answer yes:** Ryan and Frederick, "Energy, Personality, and Health," 539–541.

Part One: Cultivate Vitality

CHAPTER 1: THE PEBBLES IN YOUR SHOE

5 **at least five:** "DSM-5 Update: October 2017" (PDF), online supplement to *Diagnostic and Statistical Manual of Mental Disorders*, 5th ed. (Arlington, VA: American Psychiatric Association Publishing, 2017).

6 **an "almost diagnosis":** Jefferson Prince and Shelley Carson, *Almost Depressed: Is My (or My Loved One's) Unhappiness a Problem?* (Center City, MN: Hazelden, 2013). Prince and Carson, instructors and researchers of psychiatry and

psychology at Harvard University, describe this gray area in which a patient may not have a full clinical depression diagnosis but still not be mentally strong.

6 **Seemingly minor occurrences:** David M. Almeida, "Resilience and Vulnerability to Daily Stressors Assessed via Diary Methods," *Psychology* 14, no. 2 (April 1, 2005): 64–68, https://doi.org/10.1111/j.0963-7214.2005.00336.x.

6 **the top ten daily events:** Harvard Opinion Research Program Survey Series, *The Burden of Stress in America* (conducted by Harvard School of Public Health from March 5 to April 8, 2014, in partnership with NPR and the Robert Wood Johnson Foundation), www.rwjf.org/content/dam/farm/reports/surveys_and_polls/2014/rwjf414295.

7 **"[T]hese kinds of stressors":** Allen D. Kanner et al., "Comparison of Two Modes of Stress Measurement: Daily Hassles and Uplifts versus Major Life Events," *Journal of Behavioral Medicine* 4, no. 1 (March 1981): 1–39, https://doi.org/10.1007/BF00844845.

7 **better predictors:** Kanner et al., "Comparison of Two Modes," 1–39.

7 **record their daily microstressors:** David M. Almeida, "Daily Stressors Assessed via Diary Methods," *Current Directions in Psychological Science* 14 (2005), 64, https://pdfs.semanticscholar.org/63bf/9a0cf622e3b308b7d526d03f6ac3bcb43458.pdf?_ga=2.241262658.1921285202.1505312234-1066550283.1505312234.

7 **Watching a stressful soccer match:** Ute Wilbert-Lampen et al., "Cardiovascular Events during World Cup Soccer," *New England Journal of Medicine* 358, no. 5 (January 31, 2008): 475–83, https://doi.org/10.1056/NEJMoa0707427.

8 **Students who were about to:** L. Jeannine Petry et al., "Relationship of Stress, Distress, and the Immunologic Response to a Recombinant Hepatitis B Vaccine," *Journal of Family Practice* 32, no. 5 (May 1, 1991): 481–86, www.thefreelibrary.com/Relationship+of+stress%2c+distress%2c+and+immunologic+response+to+a . . . -a010881161.

8 **People who report:** Ronald Glaser et al., "The Influence of Psychological Stress on the Immune Response to Vaccines," *Annals of the New York Academy of Sciences* 840, no. 1 (February 7, 2006): 649–55, https://doi.org/10.1111/j.1749-6632.1998.tb09603.x.

CHAPTER 2: TIRED, STRESSED, BORED

13 **In a national survey:** Greg Toppo, "Our High School Kids: Tired, Stressed, and Bored," *USA Today*, October 23, 2015, www.usatoday.com/story/news/nation/2015/10/23/survey-students-tired-stressed-bored/74412782/.

13 **The irony, of course:** Heather Scherschel Wagner et al., "The Myth of Comfort Food," *Health Psychology* 33, no. 12 (December 2014): 1552–57, https://doi.org/10.1037/hea0000068.

13 **Academic journals are filled:** Christopher K. Hsee and Reid Hastie, "Decision and Experience: Why Don't We Choose What Makes Us Happy?," *Trends in Cognitive Sciences* 10, no. 1 (January 2006): 31–37, https://doi.org/10.1016/j.tics.2005.11.007.

14 **a study, "The Guilty Couch Potato":** Leonard Reinecke et al., "The Guilty Couch Potato: The Role of Ego Depletion in Reducing Recovery through Media Use," *Journal of Communication* 64, no. 4 (August 2014): 569–89, https://doi.org/10.1111/jcom.12107.

14 **"the doings" of their daily life:** Brian R. Little, "Well-doing: Personal Projects and the Quality of Lives," *Theory and Research in Education* 12, no. 3 (2014): 330, https://doi.org/10.1177/1477878514545847.

16 **affirming one's values:** Sander Thomaes et al., "Arousing 'Gentle Passions' in Young Adolescents: Sustained Experimental Effects of Value Affirmations on Prosocial Feelings and Behaviors," *Developmental Psychology* 48, no. 1 (January 2012): 103, https://doi.org/10.1037/a0025677.

CHAPTER 3: LITTLE *r* RESILIENCE

18 **Psychologist George A. Bonanno:** George A. Bonanno, "Loss, Trauma, and Human Resilience: Have We Underestimated the Human Capacity to Thrive After Extremely Aversive Events?," *American Psychologist* 59, no. 1 (February 2004): 20, https://doi.apa.org/doiLanding?doi=10.1037%2F0003-066X.59.1.20.

19 **These individual resilience factors:** David Palmiter et al., "Building Your Resilience," American Psychological Association, 2012, www.apa.org/helpcenter/road-resilience.aspx.

19 **more than twelve hundred people:** Carolyn M. Aldwin et al., "Do Hassles Mediate between Life Events and Mortality in Older Men?: Longitudinal Findings from the VA Normative Aging Study," *Experimental Gerontology* 59 (November 2014): 74–80, https://doi.org/10.1016/j.exger.2014.06.019.

20 **Professor Almeida and his team:** Science News, "Reactions to Everyday Stressors Predict Future Health," *ScienceDaily*, November 2, 2012, www.sciencedaily.com/releases/2012/11/121102205143.htm.

20 **people who are easily upset:** Kate A. Leger et al., "Personality and Stressor-Related Affect," *Journal of Personality and Social Psychology* 111, no. 6 (December 2016): 917–28, https://doi.org/10.1037/pspp0000083.

21 **Building vitality relies on boosting:** Richard M. Ryan and Edward L. Deci, "From Ego Depletion to Vitality: Theory and Findings Concerning the Facilitation of Energy Available to the Self," *Social and Personality Psychology Compass* 2, no. 2 (2008): 702–717, https://doi.org/10.1111/j.1751-9004.2008.00098.x.

25 **When a Harvard School:** Harvard Opinion Research Program Survey Series, *The Burden of Stress in America* (conducted by Harvard School of Public Health from March 5 to April 8, 2014, in partnership with NPR and the Robert Wood Johnson Foundation), www.rwjf.org/content/dam/farm/reports/surveys_and_polls/2014/rwjf414295.

26 **Whether introverted or extraverted:** William Fleeson et al., "An Intraindividual Process Approach to the Relationship between Extraversion and Positive Affect: Is Acting Extraverted as 'Good' as Being Extraverted?," *Journal of Personality and Social Psychology* 83, no. 6 (January 2003): 1409, https://doi.org/10.1037/0022-3514.83.6.1409.

26 **Contributing to something:** S. Katherine Nelson-Coffey et al., "Do Unto Others or Treat Yourself? The Effects of Prosocial and Self-Focused Behavior on Psychological Flourishing," *Emotion* 16, no. 6 (September 2016): 850, https://doi.org/10.1037/emo0000178.

26 **Learning something new:** Jon Clifton, "Mood of the World Upbeat on International Happiness Day," Gallup, March 19, 2005, https://news.gallup.com/poll/182009/mood-world-upbeat-international-happiness-day.aspx.

26 **Doing something creative:** Tamlin S. Conner et al., "Everyday Creative Activity as a Path to Flourishing," *Journal of Positive Psychology* 13, no. 2 (November 17, 2016): 181–89, https://doi.org/10.1080/17439760.2016.1257049.

26 **The more hours:** Jim Asplund, "When Americans Use Their Strengths More, They Stress Less," Gallup, September 27, 2012, https://news.gallup.com/poll/157679/americans-strengths-stress-less.aspx?utm_source=link_newsv9&utm_campaign=item_158573&utm_medium=copy.

26 **Everyday opportunities and activities:** Harry T. Reis et al., "Daily Well-Being: The Role of Autonomy, Competence, and Relatedness," *Personality and Social Psychology Bulletin* 26, no. 4 (2000): 419–435, https://doi.org/10.1177/0146167200266002.

CHAPTER 4: PEOPLE CHANGE

28 **Today, we know:** Gary Stix, "New Clues to Just How Much the Adult Brain Can Change," *Scientific American* Talking Back (blog), July 14, 2014, https://blogs.scientificamerican.com/talking-back/new-clues-to-just-how-much-the-adult-brain-can-change/.

29 **In 2011, researchers:** Katherine Woollett and Eleanor A. Maguire, "Acquiring 'the Knowledge' of London's Layout Drives Structural Brain Changes, *Current Biology* 21, no. 24 (December 20, 2011), 2109–114, https://doi.org/10.1016/j.cub.2011.11.018.

29 **Learning something new creates:** The paragraphs about the London cabbies were adapted from my blog: Samantha Boardman, "Your Brain Is a Muscle You Can Strengthen: Here's How," Positive Prescription (blog), accessed January 26, 2021, https://positiveprescription.com/brain-muscle-can-strengthen-heres/.

29 **A study that spanned:** Mathew A. Harris et al., "Personality Stability from Age 14 to Age 77 Years," *Psychology and Aging* 31, no. 8 (2016): 862–74, http://dx.doi.org/10.1037/pag0000133.

29 **Letting go of the notion:** Jessica Schleider and John Weisz, "A Single-Session Growth Mindset Intervention for Adolescent Anxiety and Depression: 9-Month Outcomes of a Randomized Trial," *Journal of Child Psychology and Psychiatry* 59, no. 2 (September 18, 2017): 160–70, https://doi.org/10.1111/jcpp.12811.

29 **the topic of neuroplasticity:** David S. Yeager et al., "How to Improve Adolescent Stress Responses: Insights from Integrating Implicit Theories of Personality and Biopsychosocial Models," *Psychological Science* 27, no. 8 (June 20, 2016): 1078–91, https://doi.org/10.1177/0956797616649604.

30 **Greater job satisfaction decreases neuroticism:** Christie Napa Scollon and Ed Diener, "Love, Work, and Changes in Extraversion and Neuroticism over Time," *Journal of Personality and Social Psychology* 91, no. 6 (January 2007): 1152–65, https://doi.org/10.1037/0022-3514.91.6.1152.

30 **Active people:** Yannick Stephan et al., "Physical Activity and Personality Development over Twenty Years: Evidence from Three Longitudinal Samples," *Journal of Research in Personality* 73 (April 2018): 173–79, https://doi.org/10.1016/j.jrp.2018.02.005.

30 **As journalist Lindsay Crouse:** Lindsay Crouse, "I Am 35 and Running Faster Than I Ever Thought Possible," *The New York Times*, January 31, 2020, www.nytimes.com/2020/01/31/opinion/sunday/olympic-runners-women-qualifiers.html?action=click&module=Opinion&pgtype=Homepage.

30 **It is not just through experience:** Nathan W. Hudson et al., "You Have to Follow Through: Attaining Behavioral Change Goals Predicts Volitional Personality Change," *Journal of Personality and Social Psychology* 117, no. 4 (October 2019): 839–57, https://doi.org/10.1037/pspp0000221.

31 **Students who are taught:** David S. Yeager et al., "Implicit Theories of Personality and Attributions of Hostile Intent: A Meta-Analysis, an Experiment, and a Longitudinal Intervention," *Child Development* 84, no. 5 (February 12, 2013): 1651–67, https://doi.org/10.1111/cdev.12062.

32 **Whenever you find yourself:** Douglas Starr, "The Bias Detective," *Science* 367, no. 6485 (March 27, 2020): 1418–21, https://doi.org/10.1126/science.367.6485.1418.

35 **young adults who worked concurrently:** Michael D. Mrazek et al., "Pushing the Limits: Cognitive, Affective, and Neural Plasticity Revealed by an Intensive Multifaceted Intervention," *Frontiers in Human Neuroscience* 10 (March 18, 2016): 117, https://doi.org/10.3389/fnhum.2016.00117.

36 **Lead researcher Michael Mrazek:** Science News, "Change by the Bundle: Study Shows People Are Capable of Multiple, Simultaneous Life Changes," *ScienceDaily*, March 25, 2016, www.sciencedaily.com/releases/2016/03/160325093732.htm.

Part Two: Choose Vitality

CHAPTER 5: HARD, BUT IN A GOOD WAY

40 **in a separate experiment:** L. E. Crawford et al., "Enriched Environment Exposure Accelerates Rodent Driving Skills," *Behavioural Brain Research* 378 (January 27, 2020): 112309, https://doi.org/10.1016/j.bbr.2019.112309.

40 **"a kind of mental vitamin":** Kelly Lambert, *Lifting Depression: A Neuroscientist's Hands-On Approach to Activating Your Brain's Healing Power* (New York: Basic Books, 2010), 36.

41 **Robert and Elizabeth Bjork's research:** Elizabeth L. Bjork and Robert A. Bjork, "Making Things Hard on Yourself, But in a Good Way: Creating Desirable Difficulties to Enhance Learning." *Psychology and the Real World: Essays*

Illustrating Fundamental Contributions to Society 2, no. 59–68 (2011), https://bjorklab.psych.ucla.edu/wp-content/uploads/sites/13/2016/04/EBjork_RBjork_2011.pdf.

41 **"the father of stress research":** Siang Yong Tan and A. Yip, "Hans Selye (1907–1982): Founder of the Stress Theory," *Singapore Medical Journal* 54, no. 4 (April 2018): 170–71, https://doi.org/10.11622/smedj.2018043.

42 **phenomenon the "IKEA effect":** Michael I. Norton et al., "The IKEA Effect: When Labor Leads to Love," *Journal of Consumer Psychology* 22, no. 3 (July 2012): 453–60, https://doi.org/10.1016/j.jcps.2011.08.002.

43 **state-of-the-art cake mix:** Michael Y. Park, "A History of the Cake Mix, the Invention That Redefined 'Baking'," *Bon Appétit*, September 26, 2013, www.bonappetit.com/entertaining-style/pop-culture/article/cake-mix-history. Of note, this account has been challenged within the *Harvard Business Review* paper "The Ikea Effect," which cites Laura Shapiro's book *Something from the Oven*. Also, *Bon Appétit* magazine states the real game changer was the icing: "But the innovation that saved the cake mix wasn't the egg—it was the icing on the cake."

43 **We are at risk:** Tim Wu, "The Tyranny of Convenience," *The New York Times*, February 16, 2018, www.nytimes.com/2018/02/16/opinion/sunday/tyranny-convenience.html.

44 **Being at work and:** Mathew P. White and Paul Dolan, "Accounting for the Richness of Daily Activities," *Psychological Science* 20, no. 8 (August 1, 2009): 1000–8, https://doi.org/10.1111/j.1467-9280.2009.02392.x.

45 **Psychiatrist Richard Friedman:** Richard Friedman, "Is Burnout Real?," *The New York Times*, June 3, 2019, www.nytimes.com/2019/06/03/opinion/burnout-stress.html.

46 **Poet David Whyte describes:** David Whyte, *A Lyrical Bridge between Past, Present, and Future*, filmed in 2017, TED video, 20:08, www.ted.com/talks/david_whyte_a_lyrical_bridge_between_past_present_and_future?language=en/.

46 **his book *The Happiness Hypothesis*:** Jonathan Haidt, *The Happiness Hypothesis: Finding Modern Truth in Ancient Wisdom* (New York: Basic Books, 2006), 83–84.

CHAPTER 6: BE UN-YOU

49 **A study of children:** Rachel E. White et al., "The 'Batman Effect': Improving Perseverance in Young Children," *Child Development* 88, no. 5 (September 2017): 1563–71, https://doi.org/10.1111/cdev.12695.

51 **Tapping into the capabilities:** Srini Pillay, "Your Brain Can Only Take So Much Focus," *Harvard Business Review*, May 12, 2017, https://hbr.org/2017/05/your-brain-can-only-take-so-much-focus.

51 **people demonstrated greater flexibility:** Denis Dumas and Kevin N. Dunbar, "The Creative Stereotype Effect," *PLoS ONE* 11, no. 2 (February 10, 2016): e0142567, https://doi.org/10.1371/journal.pone.0142567.

51 **expressing your "true self":** Herminia Ibarra, "The Authenticity Paradox," *Harvard Business Review*, February 2015, https://hbr.org/2015/01/the-authenticity-paradox.

51 **Believing there is a singular:** Carol S. Dweck, *Mindset: The New Psychology of Success* (New York: Ballantine Books, 2016).

51 **has been linked with depression:** Adriana Sum Miu and David Scott Yeager, "Preventing Symptoms of Depression by Teaching Adolescents That People Can Change: Effects of a Brief Incremental Theory of Personality Intervention at 9-Month Follow-Up," *Clinical Psychological Science* 3, no. 5 (September 15, 2014): 726–43, https://doi.org/10.1177/2167702614548317.

53 **she would have said yes:** Samantha Boardman, "How to Not be a People Pleaser," *Marie Claire*, June 6, 2018, www.marieclaire.com/health-fitness/a21085444/people-pleaser-advice-tips/. People pleasing can be a recipe for burnout and bitterness. Adapted from my article in *Marie Claire*.

53 **being a well-meaning phony:** Joshua Rothman, "The Art of Decision-Making," *The New Yorker*, January 14, 2019, www.newyorker.com/magazine/2019/01/21/the-art-of-decision-making. Life choices aren't just about what you want to do, they're about who you want to be. As Rothman states, "Callard maintains, we 'aspire' to self-transformation by trying on the values that we hope one day to possess."

53 **Authenticity in a relationship:** Muping Gan and Serena Chen, "Being Your Actual or Ideal Self? What It Means to Feel Authentic in a Relationship," *Personality and Social Psychology Bulletin* 43, no. 4 (February 13, 2017): 465–78, https://doi.org/10.1177/0146167216688211.

54 **Shy people feel better:** William Fleeson et al., "An Intraindividual Process Approach to the Relationship between Extraversion and Positive Affect: Is Acting Extraverted as 'Good' as Being Extraverted?," *Journal of Personality and Social Psychology* 83, no. 6 (2002): 1409, https://doi.org/10.1037/0022-3514.83.6.1409.

54 **Psychologist Sonja Lyubomirsky:** Seth Margolis and Sonja Lyubomirsky, "Experimental Manipulation of Extraverted and Introverted Behavior and Its Effects on Well-Being," *Journal of Experimental Psychology: General* 149, no. 4 (April 2019): 719–31, https://doi.org/10.1037/xge0000668.

54 **We wonder what we would:** Keith Oatley, "Fiction: Simulation of Social Worlds," *Trends in Cognitive Sciences* 20, no. 8 (August 2016): 618–28, https://doi.org/10.1016/j.tics.2016.06.002.

54 **It can also be helpful:** Jessica Black and Jennifer L. Barnes, "Fiction and Social Cognition: The Effect of Viewing Award-Winning Television Dramas on Theory of Mind," *Psychology of Aesthetics, Creativity, and the Arts* 9, no. 4 (September 2015): 423, https://doi.org/10.1037/aca0000031.

54 **On the contrary, most say:** William Fleeson and Joshua Wilt, "The Relevance of Big Five Trait Content in Behavior to Subjective Authenticity: Do High Levels of Within-Person Behavioral Variability Undermine or Enable Authenticity Achievement?," *Journal of Personality* 78, no. 4 (August 2010): 1353–82, https://doi.org/10.1111/j.1467-6494.2010.00653.x.

CHAPTER 7: EVERYONE STUMBLES

56 **Olympic figure skater:** Sarah Hughes, "Role Models Can Serve as the Ultimate Inspiration, *The New York Times*, August 16, 2016, www.nytimes.com/roomfordebate/2016/08/16/how-do-olympians-stay-so-driven/role-models-can-serve-as-the-ultimate-inspiration.

58 **Mrs. Obama's relatability:** Samantha Boardman, "The Power of Inspiration," Positive Prescription (blog), accessed January 27, 2021, https://positiveprescription.com/the-power-of-inspiration/.

58 **An MIT study found:** Julia A. Leonard et al., "Infants Make More Attempts to Achieve a Goal when They See Adults Persist," *Science* 357, no. 6357 (September 22, 2017): 1290–94, https://doi.org/10.1126/science.aan2317.

58 **the children of working mothers:** Kathleen L. McGinn et al., "Learning from Mum: Cross-National Evidence Linking Maternal Employment and Adult Children's Outcomes," *Work, Employment and Society* 33, no. 3 (April 30, 2019): 374–400, https://doi.org/10.1177/0950017018760167.

59 **the Duck syndrome:** Julie Scelfo, "Suicide on Campus and the Pressure of Perfection, *The New York Times*, July 27, 2015, www.nytimes.com/2015/08/02/education/edlife/stress-social-media-and-suicide-on-campus.html.

60 **demonstrates how misperceptions:** Alexander H. Jordan et al., "Misery Has More Company Than People Think: Underestimating the Prevalence of Others' Negative Emotions," *Personality and Social Psychology Bulletin* 37, no. 1 (January 2011): 120–35, https://doi.org/10.1177/0146167210390822.

60 **most first-year students believe:** Ashley V. Whillans et al., "From Misperception to Social Connection: Correlates and Consequences of Overestimating Others' Social Connectedness," *Personality and Social Psychology Bulletin* 43, no. 12 (September 14, 2017): 1696–711, https://doi.org/10.1177/0146167217727496.

61 **Chekhov once wrote:** Jordan et al., "Misery Has More Company," 120–135.

61 **Festinger formulated "social comparison theory":** *Social Comparison: Contemporary Theory and Research*, eds. Jerry Suls and Thomas Ashby Wills (Hillsdale, NJ: Lawrence Erlbaum Associates, 1991).

61 **Research suggests that unhappy people:** Sonja Lyubomirsky and Lee Ross, "Hedonic Consequences of Social Comparison: A Contrast of Happy and Unhappy People," *Journal of Personality and Social Psychology* 73, no. 6 (December 1997): 1141, https://doi.org/10.1037/0022-3514.73.6.1141.

61 **The tendency to seek:** Judith B. White et al., "Frequent Social Comparisons and Destructive Emotions and Behaviors: The Dark Side of Social Comparisons," *Journal of Adult Development* 13, no. 1 (June 14, 2006): 36–44, https://doi.org/10.1007/s10804-006-9005-0.

62 **the performative side of Instagram:** Grace Holland and Marika Tiggemann, "A Systematic Review of the Impact of the Use of Social Networking Sites on Body Image and Disordered Eating Outcomes," *Body Image* 17 (June, 2016): 100–110, https://doi.org/10.1016/j.bodyim.2016.02.008; Marika Tiggemann and Belinda McGill, "The Role of Social Comparison in the Effect of Magazine

Advertisements on Women's Mood and Body Dissatisfaction," *Journal of Social and Clinical Psychology* 23, no. 1 (February 2004): 23–44, https://doi.org/10.1521/jscp.23.1.23.26991; Jacqueline V. Hogue and Jennifer S. Mills, "The Effects of Active Social Media Engagement with Peers on Body Image in Young Women," *Body Image* 28 (March 2019): 1–5, https://doi.org/10.1016/j.bodyim.2018.11.002; Marika Tiggemann et al., "The Processing of Thin Ideals in Fashion Magazines: A Source of Social Comparison or Fantasy?," *Journal of Social and Clinical Psychology* 28, no. 1 (January 2009): 73–93, https://doi.org/10.1521/jscp.2009.28.1.73; Marika Tiggemann and Isabella Barbato, "You Look Great!": The Effect of Viewing Appearance-Related Instagram Comments on Women's Body Image," *Body Image* 27 (December 2018): 61–66, https://doi.org/10.1016/j.bodyim.2018.08.009; and Marika Tiggemann and Isabella Anderberg, "Social Media Is Not Real: The Effect of 'Instagram vs Reality' Images on Women's Social Comparison and Body Image," *New Media & Society* 22, no. 12 (December 1, 2020): 2183–99, https://doi.org/1461444819888720.

CHAPTER 8: BETTER DAYS

63 **Pathogenesis—the treatment of disease:** Aaron Antonovsky, "The Salutogenic Model as a Theory to Guide Health Promotion," *Health Promotion International* 11, no. 1 (March 1996): 11–18, https://doi.org/10.1093/heapro/11.1.11.

63 **We are bombarded:** "The Grand Experiment: Are Our Brains Outsmarting Themselves? A Richmond Researcher and Her Rats Say Maybe," *University of Richmond Magazine*, May 1, 2018, https://magazine.richmond.edu/features/article/-/15317/the-grand-experiment.html. In an interview in *University of Richmond Magazine*, professor of behavioral neuroscience Kelly Lambert discusses the endless choices we have to make each day. A significant focus of Lambert's work has been stress and adaptive resilience, or the ability to positively respond to the stressors and uncertainties that life throws at us. From the simple "Do you want fries with that?" to the life-changing "Will you marry me?," the inconsequential selection of a shirt in the morning to the momentous commitment to have a child, we are barraged every day with the endless demands of choices to be made and actions to be taken. "Our lives," says Lambert, "are just one decision after another."

63 **Confidence in our ability:** Seunghee Han et al., "Feelings and Consumer Decision Making: The Appraisal-Tendency Framework," *Journal of Consumer Psychology* 17, no. 3 (January 22, 2008): 158–68, https://doi.org/10.1016/S1057-7408(07)70023-2.

63 **results of one study indicated:** Francesca Gino et al., "Anxiety, Advice, and the Ability to Discern: Feeling Anxious Motivates Individuals to Seek and Use Advice," *Journal of Personality and Social Psychology* 102, no. 3 (March 2012): 497, https://doi.org/10.1037/a0026413.

64 **The accumulation of wealth:** L. Parker Schiffer and Tomi-Ann Roberts, "The Paradox of Happiness: Why Are We Not Doing What We Know Makes Us

Happy?," *Journal of Positive Psychology* 13, no. 3 (January 11, 2017): 252–59, https://doi.org/10.1080/17439760.2017.1279209.

64 **We quickly adapt:** Ed O'Brien and Samantha Kassirer, "People Are Slow to Adapt to the Warm Glow of Giving," *Psychological Science* 30, no. 2 (February 1, 2019): 193–204, https://doi.org/10.1177/0956797618814145.

64 **Low-effort activities:** Chen Zhang et al., "More Is Less: Learning But Not Relaxing Buffers Deviance Under Job Stressors," *Journal of Applied Psychology* 103, no. 2 (February 2018): 123, https://doi.org/10.1037/apl0000264.

64 **They appeal to our instinct:** Christopher K. Hsee et al., "Idleness Aversion and the Need for Justifiable Busyness," *Psychological Science* 21, no. 7 (July 1, 2010): 926–30, https://doi.org/10.1177/0956797610374738.

65 **Everyday well-being resides:** Tina Söderbacka et al., "Older Persons' Experiences of What Influences Their Vitality—A Study of 65- and 75-Year-Olds in Finland and Sweden," *Scandinavian Journal of Caring Sciences* 31, no. 2 (June 2017): 378–87, https://doi.org/10.1111/scs.12357. Vitality is typically mentioned in the context of the elderly but has application across the lifespan.

65 **When asked what wellness means:** Megan Thielking, "Turning the Tables, People with Mental Illness Share What They Want Scientists to Study," STAT, October 24, 2018, www.statnews.com/2018/10/24/mental-illness-survey-asks-what-scientists-should-study/. The study discussed in this piece was conducted by the Milken Institute and the Depression and Bipolar Support Alliance.

65 **They showed how uplifts counteract:** Anita DeLongis et al., "Relationship of Daily Hassles, Uplifts, and Major Life Events to Health Status," *Health Psychology* 1, no. 2 (1982): 119, https://delongis-psych.sites.olt.ubc.ca/files/2018/03/Relationship-of-Daily-Hassles.pdf.

65 **Today there is increasing evidence:** Kristin Layous et al., "Delivering Happiness: Translating Positive Psychology Intervention Research for Treating Major and Minor Depressive Disorders," *Journal of Alternative and Complementary Medicine* 17, no. 8 (August 2011): 675–83, https://doi.org/10.1089/acm.2011.0139. Contrasted with the narrowing of attention and behavioral inhibition characteristic of negative states, "positive emotions trigger upward spirals toward greater flourishing, resilience, and psychologic well-being."

65 **By generating positive emotions:** Kristin Layous et al., "Positive Activities As Protective Factors against Mental Health Conditions," *Journal of Abnormal Psychology* 123, no. 1 (February 2014): 3, https://doi.org/10.1037/a0034709.

66 **Importantly, positive emotions:** Susan Folkman and Judith Tedlie Moskowitz, "Stress, Positive Emotion, and Coping," *Current Directions in Psychological Science* 9, no. 4 (August 1, 2000): 115–18, https://doi.org/10.1111/1467-8721.00073. People can experience positive emotions in the midst of difficult and even painful circumstances. In the late 1990s, researchers Susan Folkman and Judith Moskowitz monitored men who were the primary caregivers of partners with AIDS. Positive moments helped the caregivers manage their stress and facilitated coping: "Month after month, more than 99 percent of the caregivers noted and remembered positive events in the midst of some of the most psychologically

stressful circumstances people encounter." The events were ordinary events of daily life—receiving a compliment for something minor, preparing a meal with friends, noticing something beautiful. The caregivers deliberately noted them, created them, and also infused them with meaning.

66 **Positive emotions can "undo":** Michele M. Tugade et al., "Psychological Resilience and Positive Emotional Granularity: Examining the Benefits of Positive Emotions on Coping and Health," *Journal of Personality* 72, no. 6 (December 2004): 1161–90, https://doi.org/10.1111/j.1467-6494.2004.00294.x.

66 **less likely to catch a cold:** Sheldon Cohen et al., "Positive Emotional Style Predicts Resistance to Illness After Experimental Exposure to Rhinovirus or Influenza A Virus," *Psychosomatic Medicine* 68, no. 6 (December 2006): 809–15, https://doi.org/10.1097/01.psy.0000245867.92364.3c and Andrew Steptoe et al., "Neuroendocrine and Inflammatory Factors Associated with Positive Affect in Healthy Men and Women: The Whitehall II Study," *American Journal of Epidemiology* 167, no. 1 (October 4, 2007): 96–102, https://doi.org/10.1093/aje/kwm252.

67 **Arguments with loved ones:** Corinna Wu, "Sweating the Small Stuff," *ASEE Prism* 14, no. 2 (October 2004): 22, www.jstor.org/stable/i24161150.

67 **interpret a bother as a challenge:** Vikas Mittal and William T. Ross Jr., "The Impact of Positive and Negative Affect and Issue Framing on Issue Interpretation and Risk Taking," *Organizational Behavior and Human Decision Processes* 76, no. 3 (December 1998): 298–324, https://doi.org/10.1006/obhd.1998.2808.

67 **An uplift can help reduce:** Michael A. Cohn et al., "Happiness Unpacked: Positive Emotions Increase Life Satisfaction by Building Resilience," *Emotion* 9, no. 3 (2009): 361, https://doi.org/10.1037/a0015952.

67 **Although they may be short lived:** Barbara L. Fredrickson and Christine Branigan, "Positive Emotions Broaden the Scope of Attention and Thought-Action Repertoires," *Cognition and Emotion* 19, no. 3 (June 20, 2011): 313–32, https://doi.org/10.1080/02699930441000238.

67 **They build social, intellectual:** Barbara L. Fredrickson and Thomas Joiner, "Positive Emotions Trigger Upward Spirals toward Emotional Well-Being." *Psychological Science* 13, no. 2 (2002): 172–5.

67 **"upward spiral" of growth:** Michele M. Tugade and Barbara L. Fredrickson, "Resilient Individuals Use Positive Emotions to Bounce Back from Negative Emotional Experiences," *Journal of Personality and Social Psychology* 86, no. 2 (2004): 320, www.ncbi.nlm.nih.gov/pmc/articles/PMC3132556/.

68 **Expanded thinking and attention:** Kristin Layous et al., "Positive Activities as Protective Factors against Mental Health Conditions," *Journal of Abnormal Psychology* 123, no. 1 (February 2014): 3, https://doi.org/10.1037/a0034709.

68 **These brief "lived experiences":** Cohn et al., "Happiness Unpacked," 361. "Unlike negative emotions, which narrow attention, cognition, and physiology toward coping with an immediate threat or problem (Cosmides & Tooby, 2000; Carver, 2003), positive emotions produce novel and broad-ranging thoughts and actions that are usually not critical to one's immediate safety, well-being, or survival."

69 **action-oriented and other-oriented:** Anne Arewasikporn et al., "Sharing Positive Experiences Boosts Resilient Thinking: Everyday Benefits of Social Connection and Positive Emotion in a Community Sample," *American Journal of Community Psychology* 63, no. 1–2 (2019): 110–121.

69 **It is especially difficult:** Daryl B. O'Connor et al., "Effects of Daily Hassles and Eating Style on Eating Behavior," *Health Psychology* 27, no. 1S (January 2008): S20, https://doi.org/10.1037/0278-6133.27.1.S20. "To conclude, the results of this research showed daily hassles were associated with a shift in preference toward high fat and high sugar between-meal snack foods and with a reduction in main meals and vegetable consumption. Ego-threatening, interpersonal and work-related hassles were significantly associated with increased snacking, whereas, physical stressors were significantly associated with decreased snacking. Daily hassles were associated with increased consumption of high fat and high sugar between-meal snack foods and also with a perceived reduction in main meals and vegetable consumption. These results are concerning as an overwhelming body of evidence has shown the importance of maintaining a balanced diet, including eating a low fat diet and five portions of fruit and vegetables a day, in terms of reducing risk of cardiovascular disease and cancer risk (e.g., Heimendinger, Van Ryn, Chapelsky, Forester, & Stables, 1996; Kumanyika et al., 2000; Van Horn & Kavey, 1997; Wong & Lam, 1999). Therefore, the changes observed in the current study may indicate a serious indirect pathway through which stress influences health risk."

69 **Instead of trying in vain:** Leah Dickens and David DeSteno, "The Grateful Are Patient: Heightened Daily Gratitude Is Associated with Attenuated Temporal Discounting," *Emotion* 16, no. 4 (June 2016): 421, https://doi.org/10.1037/emo0000176.

69 **people who engage in prosocial actions:** Kurt Gray, "Moral Transformation: Good and Evil Turn the Weak Into the Mighty," *Social Psychological and Personality Science* 1, no. 3 (July 19, 2010): 253–58, https://doi.org/10.1177/1948550610367686.

69 **As researcher Kurt Gray suggests:** Sam Savage, "Being Naughty or Nice May Boost Willpower, Physical Endurance," RedOrbit, April 19, 2010, www.redorbit.com/news/health/1851940/being_naughty_or_nice_may_boost_willpower_physical_endurance/.

69 **Walter Mischel's famous marshmallow study:** W. Mischel et al., "Cognitive and Attentional Mechanisms in Delay of Gratification," *Journal of Personality and Social Psychology* 21, no. 2 (February 1972): 204, https://doi.org/10.1037/h0032198.

70 **revealed eye-opening data:** B. J. Casey et al., "Behavioral and Neural Correlates of Delay of Gratification 40 Years Later," *Proceedings of the National Academy of Sciences* 108, no. 36 (September 6, 2011): 14998–15003, https://doi.org/10.1073/pnas.1108561108.

70 **Author Pamela Druckerman:** Pamela Druckerman, "Learning How to Exert Self-Control," *The New York Times*, September 12, 2014, www.nytimes.com

/2014/09/14/opinion/sunday/learning-self-control.html?_r=0&login=smart lock&auth=login-smartlock.

70 **recent marshmallow-related news:** Rebecca Koomen, Sebastian Grueneisen, and Esther Herrmann, "What a New Marshmallow Test Teaches Us About Cooperation," *Behaviorial Scientist* (August 17, 2020), https://behavioralscien tist.org/what-a-new-marshmallow-test-teaches-us-about-cooperation.

CHAPTER 9: MORE LIFE

71 **When study researchers asked nurses:** Joyce E. Bono et al., "Building Positive Resources: Effects of Positive Events and Positive Reflection on Work Stress and Health," *Academy of Management Journal* 56, no. 6 (December 1, 2013): 1601–27, https://doi.org/10.5465/amj.2011.0272.

71 **Doing so can help you capitalize:** Joyce E. Bono and Theresa M. Glomb, "The Powerful Effect of Noticing Good Things at Work," *Harvard Business Review*, September 4, 2015, https://hbr.org/2015/09/the-powerful-effect-of-noticing -good-things-at-work.

72 **This probably has an evolutionary basis:** Roy F. Baumeister et al., "Bad Is Stronger Than Good," *Review of General Psychology* 5, no. 4 (December 1, 2001): 323–70, https://doi.org/10.1037/1089-2680.5.4.323.

73 **The phenomenon of unfinished business:** Bluma Zeigarnik, "On Finished and Unfinished Tasks," *A Source Book of Gestalt Psychology* 1 (1938): 300–314, https:// doi.org/10.1037/11496-025.

74 **Productivity porn offers a smorgasbord:** Daniel McGinn, "Still Trying to Get More Done," *Harvard Business Review*, April 2016, https://hbr.org/2016/04 /still-trying-to-get-more-done.

74 **they hoped to make progress:** Nicole L. Mead et al., "Simple Pleasures, Small Annoyances, and Goal Progress in Daily Life," *Journal of the Association for Consumer Research* 1, no. 4 (August 2016): 527–39, https://doi.org/10.1086/688287. A "simple pleasure" is conceptualized as a brief, positive experience that emerges in everyday settings and is readily accessible to most individuals at little or no cost.

74 **A few minutes of friendly conversation:** Oscar Ybarra et al., "Friends (and Sometimes Enemies) with Cognitive Benefits: What Types of Social Interactions Boost Executive Functioning?," *Social Psychological and Personality Science* 2, no. 3 (May 1, 2011): 253–61, https://doi.org/10.1177/1948550610386808.

75 **Compared with people:** Robert A. Emmons and Michael E. McCullough, "Counting Blessings versus Burdens: An Experimental Investigation of Gratitude and Subjective Well-Being in Daily Life," *Journal of Personality and Social Psychology* 84, no. 2 (March 2003): 377, https://doi.org/10.1037/0022 -3514.84.2.377.

CHAPTER 10: TAKE ACTION

79 **A study in the *Lancet*:** David A. Richards et al., "Cost and Outcome of Behavioural Activation versus Cognitive Behavioural Therapy for Depression (COBRA): A Randomised, Controlled, Non-Inferiority Trial," *Lancet* 388, no. 10047 (September 2016): 871–80, https://doi.org/10.1016/S0140-6736(16)31140-0.

79 **BA therapy is as effective:** Science News, "Behavioral Activation as Effective as CBT for Depression, at Lower Cost," *ScienceDaily*, July 22, 2016, www.sciencedaily.com/releases/2016/07/160722212245.htm.

79 **Another study of 241 depressed patients:** Sona Dimidjian et al., "Randomized Trial of Behavioral Activation, Cognitive Therapy, and Antidepressant Medication in the Acute Treatment of Adults with Major Depression," *Journal of Consulting and Clinical Psychology* 74, no. 4 (August 2006): 658, https://doi.org/10.1037/0022-006X.74.4.658.

79 **Brain scans revealed:** S. Shiota et al., "Effects of Behavioural Activation on the Neural Basis of Other Perspective Self-Referential Processing in Subthreshold Depression: A Functional Magnetic Resonance Imaging Study," *Psychological Medicine* 47, no. 5 (April 2017): 877–88, https://doi.org/10.1017/S0033291716002956.

80 **Couples who make:** Galena Rhoades and Scott Stanley, *Before 'I Do': What Do Premarital Experiences Have to Do with Marital Quality among Today's Young Adults?* (Charlottesville, VA: National Marriage Project, 2014), http://nationalmarriageproject.org/wordpress/wp-content/uploads/2014/08/NMP-BeforeIDoReport-Final.pdf.

80 **A passive existence is not:** Jonathan W. Kanter and Ajeng J. Puspitasari, "Global Dissemination and Implementation of Behavioural Activation," *Lancet* 388, no. 10047 (August 27, 2016): 843–44, https://doi.org/10.1016/S0140-6736(16)31131-X.

80 **In a study of obese women:** Gabriele Oettingen, "Future Thought and Behaviour Change," *European Review of Social Psychology* 23, no. 1 (March 13, 2012): 1–63, https://doi.org/10.1080/10463283.2011.643698.

81 **Based on her research:** Gabriele Oettingen, *Rethinking Positive Thinking: Inside the New Science of Motivation* (New York: Current, 2015).

82 **known as implementation intention:** Martin S. Hagger and Aleksandra Luszczynska, "Implementation Intention and Action Planning Interventions in Health Contexts: State of the Research and Proposals for the Way Forward," *Applied Psychology: Health and Well-Being* 6, no. 1 (March 2014): 1–47, https://doi.org/ 10.1111/aphw.12017.

82 **Mental contrasting has been found:** Gertraud Stadler et al., "Physical Activity in Women: Effects of a Self-Regulation Intervention," *American Journal of Preventive Medicine* 36, no. 1 (January 2009): 29–34, https://doi.org/10.1016/j.amepre.2008.09.021.

82 **Students improved their grades:** Angela Lee Duckworth et al, "Self-Regulation Strategies Improve Self-Discipline in Adolescents: Benefits of Mental Con-

trasting and Implementation Intentions," *Educational Psychology* 31, no. 1 (September 14, 2010): 17–26, https://doi.org/10.1080/01443410.2010.506003.

82 **Nurses who performed:** Peter M. Gollwitzer et al., "Promoting the Self-Regulation of Stress in Health Care Providers: An Internet-Based Intervention," *Frontiers in Psychology* 9 (June 15, 2018): 838, https://doi.org/10.3389/fpsyg.2018.00838.

83 **Proactively orienting yourself:** Sylviane Houssais et al., "Using Mental Contrasting with Implementation Intentions to Self-Regulate Insecurity-Based Behaviors in Relationships," *Motivation and Emotion* 37, no. 2 (June 13, 2013): 224–33, https://doi.org/10.1007/s11031-012-9307-4.

85 **Two monks were walking:** This story may also be found at the following resource: Posted by ahlhalau, "Two Monks and a Woman—A Zen Lesson," *KindSpring*, accessed February 13, 2021, www.kindspring.org/story/view.php?sid=63753#sthash.4z5n3dsK.dpuf.

86 **The most persistent regrets:** Ian M. Davison and Aidan Feeney, "Regret as Autobiographical Memory," *Cognitive Psychology* 57, no. 4 (December 2008): 385–403, https://doi.org/10.1016/j.cogpsych.2008.03.001.

Part Three: Connect with Others

CHAPTER 11: SKIP THE CASSEROLES

91 **There is little comparable help:** Gordon L. Flett et al., "Social Support and Help-Seeking in Daily Hassles versus Major Life Events Stress," *Journal of Applied Social Psychology* 25, no. 1 (January 1995): 49–58, https://doi.org/10.1111/j.1559-1816.1995.tb01583.x.

93 **It's central:** Ed Diener and Shigehiro Oishi, "The Nonobvious Social Psychology of Happiness," *Psychological Inquiry* 16, no. 4 (November 19, 2009), 162–67, https://doi.org/10.1207/s15327965pli1604_04 and John F. Helliwell and Robert D. Putnam, "The Social Context of Well-Being," *Philosophical Transactions of the Royal Society B* 359 (September 29, 2004): 1435–46, https://doi.org/10.1098/rstb.2004.1522.

93 **I told Suzanne about a study:** Julia M. Rohrer et al., "Successfully Striving for Happiness: Socially Engaged Pursuits Predict Increases in Life Satisfaction," *Psychological Science* 29, no. 8 (August 1, 2018): 1291–98, https://doi.org/10.1177/0956797618761660.

94 **Similarly, exercise that involves:** Sammi R. Chekroud et al., "Association between Physical Exercise and Mental Health in 1.2 Million Individuals in the USA between 2011 and 2015: A Cross-Sectional Study," *Lancet Psychiatry* 5, no. 9 (September 2018): 739–46, https://doi.org/10.1016/S2215-0366(18)30227-X.

94 **a study of Danish men and women:** Peter Schnohr et al., "Various Leisure-Time Physical Activities Associated with Widely Divergent Life Expectancies: The Copenhagen City Heart Study," *Mayo Clinic Proceedings* 93, no. 12 (December 1, 2018): 1775–85, https://doi.org/10.1016/j.mayocp.2018.06.025.

94 **Behavioral scientists call this:** Todd Rogers et al., "Commitment Devices: Using Initiatives to Change Behavior," *Journal of American Medical Association* 311, no. 20 (May 18, 2014): 2065–66, https://doi.org/10.1001/jama.2014.3485.

95 **Lacking positive relationships:** Julianne Holt-Lunstad and Timothy B. Smith, "Social Relationships and Mortality," *Social and Personality Psychology Compass* 6, no. 1 (January 3, 2012): 41–53, https://doi.org/10.1111/j.1751-9004.2011.00406.x.

95 **an individual's personal qualities:** David S. Lee et al., "I-through-We: How Supportive Social Relationships Facilitate Personal Growth," *Personality and Social Psychology Bulletin* 44, no. 1 (January 1, 2018): 37–48, https://doi.org/10.1177/0146167217730371. "A person's growth depends not only on individual capabilities but also on his or her relational network and social capital. There is no personal growth without the individual, but growth is embedded in a social context that facilitates a person's relevant attitudes and capacities."

95 **A singular focus:** Lee et al., "I-through-We,"39.

96 **They are partners who help:** Christian Swann et al., "Exploring the Interactions Underlying Flow States: A Connecting Analysis of Flow Occurrence in European Tour Golfers," *Psychology of Sport and Exercise* 16, pt. 3 (March 2015): 60–69, https://doi.org/10.1016/j.psychsport.2014.09.007.

96 **Vermeer was deeply engaged:** Books and Arts, "Vermeer Was Brilliant, But He Was Not without Influences," *The Economist*, October 14, 2017, www.economist.com/books-and-arts/2017/10/12/vermeer-was-brilliant-but-he-was-not-without-influences.

96 **Psychologists at the University of Michigan:** Lee et al., "I-through-We," 37–48.

97 **The best relationships serve:** Brooke C. Feeney and Nancy L. Collins, "A New Look at Social Support: A Theoretical Perspective on Thriving through Relationships," *Personality and Social Psychology Review* 19, no. 2 (August 14, 2014): 113–47, https://doi.org/10.1177/1088868314544222.

97 **Studies show that a hill:** Simone Schnall et al., "Social Support and the Perception of Geographical Slant," *Journal of Experimental Social Psychology* 44, no. 5 (September 2008): 1246–55, https://doi.org/10.1016/j.jesp.2008.04.011.

97 **Public speaking seems less stressful:** Karen M. Grewen et al., "Warm Partner Contact Is Related to Lower Cardiovascular Reactivity," *Behavioral Medicine* 29, no. 3 (March 2010): 123–30, https://doi.org/10.1080/08964280309596065.

97 **a study of paramedics found:** Jessie Pow et al., "Does Social Support Buffer the Effects of Occupational Stress on Sleep Quality among Paramedics? A Daily Diary Study," *Journal of Occupational Health Psychology* 22, no. 1 (February 11, 2017): 71, https://doi.org/10.1037/a0040107.

97 **another study of married couples:** James A. Coan et al., "Lending a Hand: Social Regulation of the Neural Response to Threat," *Psychological Science* 17, no. 12 (December 2006): 1032–39, https://doi.org/10.1111/j.1467-9280.2006.01832.x. Loved ones functioned as better "emotion regulators" than strangers. As the researchers observed, "Both stranger and spousal hand-holding appear capable of decreasing the need for a coordinated bodily response to threatening stimuli, but

only spousal hand-holding confers the additional benefit of decreasing the need for vigilance, evaluation, and self-regulation of affect."

97 **Hand-holding reduced:** Ellen Berscheid, "The Human's Greatest Strength: Other Humans," in *A Psychology of Human Strengths: Fundamental Questions and Future Directions for a Positive Psychology*, eds. L. G. Aspinwall and U. M. Staudinger (Washington, DC: American Psychological Association, 2003), chap. 3, https://doi.org/10.1037/10566-003 and J. C. Coyne et al., "Prognostic Importance of Marital Quality for Survival of Congestive Heart Failure," *American Journal of Cardiology* 88, no. 5 (September 1, 2001): 526–29, https://doi.org/10.1016/S0002-9149(01)01731-3.

98 **Knowing you are loved:** Feeney and Collins, "A New Look at Social Support," 113–47.

CHAPTER 12: GAIN SOME DISTANCE

99 **When you max out:** Leonard L. Martin and Abraham Tesser, "Toward a Motivational and Structural Theory of Ruminative Thought," in *Unintended Thought*, eds. James S. Uleman and John A. Bargh (New York: Guilford Press, 1989), 306–26 and Leonard L. Martin et al., "Rumination as a Function of Goal Progress, Stop Rules, and Cerebral Lateralization," in *Depressive Rumination: Nature, Theory, and Treatment*, eds. Costas Papageorgiou and Adrian Wells (Chichester, United Kingdom: Wiley, 2004): 153–75.

99 **Spiraling inward, going over:** Jeannette M. Smith and Lauren B. Alloy, "A Roadmap to Rumination: A Review of the Definition, Assessment, and Conceptualization of This Multifaceted Construct," *Clinical Psychology Review* 29, no. 2 (March 2009): 116–28, https://doi.org/10.1016/j.cpr.2008.10.003.

100 **Participants in one study:** Ethan Kross and Ozlem Ayduk, "Facilitating Adaptive Emotional Analysis: Distinguishing Distanced-Analysis of Depressive Experiences from Immersed-Analysis and Distraction," *Personality and Social Psychology Bulletin* 34, no. 7 (July 1, 2008): 924–38, https://doi.org/10.1177/0146167208315938.

102 **study, "Flies on the Wall Are Less Aggressive":** Dominik Mischkowski et al., "Flies on the Wall Are Less Aggressive: Self-Distancing 'In the Heat of the Moment' Reduces Aggressive Thoughts, Angry Feelings, and Aggressive Behavior," *Journal of Experimental Social Psychology* 48, no. 5 (September 2012): 1187–91, https://doi.org/10.1016/j.jesp.2012.03.012.

102 **"The worst thing to do":** Science News, "'Self-Distancing' Can Help People Calm Aggressive Reactions, Study Finds," *ScienceDaily*, July 2, 2012, www.sciencedaily.com/releases/2012/07/120702153216.htm.

103 **Recognizing the transitory nature:** Emma Bruehlman-Senecal and Ozlem Ayduk, "This Too Shall Pass: Temporal Distance and the Regulation of Emotional Distress," *Journal of Personality and Social Psychology* 108, no. 2 (February 2015): 356, https://doi.org/10.1037/a0038324.

103 **Projecting into the future:** Alex C. Huynh et al., "The Value of Prospective Reasoning for Close Relationships," *Social Psychological and Personality Science* 7, no. 8 (November 1, 2016): 893–902, https://doi.org/10.1177/1948550616660591.

105 **Considering how you would advise:** Igor Grossmann and Ethan Kross, "Exploring Solomon's Paradox: Self-Distancing Eliminates the Self-Other Asymmetry in Wise Reasoning about Close Relationships in Younger and Older Adults," *Psychological Science* 25, no. 8 (August 1, 2014): 1571–80, https://doi .org/10.1177/0956797614535400.

106 **Excessive complaining and rehashing:** Amanda J. Rose et al., "Co-Rumination Exacerbates Stress Generation among Adolescents with Depressive Symptoms," *Journal of Abnormal Child Psychology* 45, no. 5 (July 2017): 985–95, https://doi .org/10.1007/s10802-016-0205-1.

107 **adolescents who co-ruminated:** Erika M. Waller and Amanda J. Rose, "Brief Report: Adolescents' Co-Rumination with Mothers, Co-Rumination with Friends, and Internalizing Symptoms," *Journal of Adolescence* 36, no. 2 (February 9, 2013): 429–33, https://doi.org/ 10.1016/j.adolescence.2012.12.006.

107 **Undergraduates who co-ruminated:** Christine A. Calmes and John E. Roberts, "Rumination in Interpersonal Relationships: Does Co-Rumination Explain Gender Differences in Emotional Distress and Relationship Satisfaction among College Students?," *Cognitive Therapy and Research* 32, no. 4 (August 2008): 577–90, https://doi.org/10.1007/s10608-008-9200-3.

108 **a short chat:** Kate J. Zelic et al., "An Experimental Investigation of Co-rumination, Problem Solving, and Distraction," *Behavior Therapy* 48, no. 3 (May 2017): 403–412, https://doi.org/10.1016/j.beth.2016.11.013.

108 **Instead of making a teenager's problems:** Marshall Duke et al., "Knowledge of Family History as a Clinically Useful Index of Psychological Well-Being and Prognosis: A Brief Report," *Psychotherapy* 45, no. 2 (June 2008): 268–72, https:// doi.org/10.1037/0033-3204.45.2.268.

109 **When you know the arc:** Rudyard Kipling, "If," Poetry Foundation, accessed January 28, 2021, www.poetryfoundation.org/poems/46473/if.

CHAPTER 13: TAILWINDS

110 **Human Beings are made up:** Tanya Luhrmann, "Beyond the Brain," *Wilson Quarterly*, Summer 2012, http://archive.wilsonquarterly.com/essays/beyond-brain.

111 **the emotions they uncover reveal:** Paul L. Wachtel, "Knowing Oneself from the Inside Out, Knowing Oneself from the Outside In: The 'Inner' and 'Outer' Worlds and Their Link through Action," *Psychoanalytic Psychology* 26, no. 2 (April 2009): 158, https://doi.org/10.1037/a0015502.

112 **study, "The Need to Belong":** Roy F. Baumeister and Mark R. Leary, "The Need to Belong: Desire for Interpersonal Attachments as a Fundamental Human Motivation," *Psychological Bulletin* 117, no. 3 (June 1995): 497–529, https://doi .org/10.1037/0033-2909.117.3.497.

112 **two in five Americans:** "New Cigna Study Reveals Loneliness at Epidemic Levels in America," Cigna, May 1, 2018, www.cigna.com/newsroom/news-releases/2018/new-cigna-study-reveals-loneliness-at-epidemic-levels-in-america.

113 **Married couples who lived apart:** Baumeister and Leary, "The Need to Belong," 512.

113 **Indeed, such "insubstantial" interactions:** Baumeister and Leary, "The Need to Belong," 512.

113 **the UK Campaign to End Loneliness:** "The Facts on Loneliness," UK Campaign to End Loneliness, accessed January 28, 2021, www.campaigntoendloneliness.org/the-facts-on-loneliness/

113 **There is some evidence:** Niall Bolger and David Amarel, "Effects of Social Support Visibility on Adjustment to Stress: Experimental Evidence," *Journal of Personality and Social Psychology* 92, no. 3 (March 2007): 458, https://doi.org/10.1037/0022-3514.92.3.458.

115 **Using a daily diary:** Niall Bolger et al, "Invisible Support and Adjustment to Stress," *Journal of Personality and Social Psychology* 79, no. 6 (December 2000): 953, https://doi.org/10.1037//0022-3514.79.6.953.

115 **During the beginning of a relationship:** Samantha Boardman, "Emotional Viagra," Positive Prescription (blog), accessed January 28, 2021, https://positiveprescription.com/emotional-viagra/.

116 **a study conducted in Israel:** G. E. Birnbaum et al., "Intimately Connected: The Importance of Partner Responsiveness for Experiencing Sexual Desire," *Journal of Personality and Social Psychology* 111, no. 4 (July 2016): 530, https://doi.org/10.1037/pspi0000069.

116 **"Being responsive is":** Science News, "Come On Baby, (Re)Light My Fire," *ScienceDaily*, July 20, 2016, www.sciencedaily.com/releases/2016/07/160720215315.htm.

117 **Sensing that your romantic partner:** Amie Gordon and Serena Chen, "Do You Get Where I'm Coming From?: Perceived Understanding Buffers Against the Negative Impact of Conflict on Relationship Satisfaction," *Journal of Personality and Social Psychology* 110, no. 2 (November 2, 2016): 239, https://doi.org/10.1037/pspi0000039.

CHAPTER 14: BETTER CONVERSATIONS

122 **"Thank you, Helen":** Samantha Boardman, "Marry Someone Smarter Than You Are," Positive Prescription (blog), accessed January 28, 2021, https://positiveprescription.com/marry-someone-smarter/.

122 **People who engage frequently:** Anne Milek et al., "'Eavesdropping on Happiness' Revisited: A Pooled, Multisample Replication of the Association between Life Satisfaction and Observed Daily Conversation Quantity and Quality," *Psychological Science* 29, no. 9 (July 2018): 1451–62, https://doi.org/10.1177/0956797618774252.

122 **"experimentally 'prescribe' people":** Science News, "Small Talk Not as Bad as Previously Thought," *ScienceDaily*, July 3, 2018, www.sciencedaily.com/releases/2018/07/180703131432.htm.

122 **Approximately half of Americans:** "New Cigna Study Reveals Loneliness at Epidemic Levels in America," Cigna, May 1, 2018, www.cigna.com/newsroom/news-releases/2018/new-cigna-study-reveals-loneliness-at-epidemic-levels-in-america.

122 **Brain scans reveal:** Diana I. Tamir and Jason P. Mitchell, "Disclosing Information about the Self Is Intrinsically Rewarding," *Proceedings of the National Academy of Sciences* 109, no. 21 (May 22, 2012): 8038–43, https://doi.org/10.1073/pnas.1202129109.

122 **People who do ask questions:** K. Huang et al., "It Doesn't Hurt to Ask: Question-Asking Increases Liking," *Journal of Personality and Social Psychology* 113, no. 3 (September 2017): 430, https://doi.org/10.1037/pspi0000097.

123 **Researchers found that when:** Huang et al., "It Doesn't Hurt to Ask," 430–32.

124 **"Conversations based on ordinary topics":** Gus Cooney, "The Unexpected Costs of Extraordinary Experiences," *Psychology Today* Real Talk (blog), May 17, 2017, www.psychologytoday.com/us/blog/real-talk/201705/the-unexpected-costs-extraordinary-experiences.

126 **intelligence researcher James Flynn:** James Flynn, *Does Your Family Make You Smarter?: Nature, Nurture, and Human Autonomy* (New York: Cambridge University Press, 2016).

CHAPTER 15: LOOK AROUND

128 **spectacle, "Musée des Beaux Arts":** W. H. Auden, "Musée des Beaux Arts," Poetry Daily, accessed January 28, 2021, https://poems.com/poem/musee-des-beaux-arts/.

128 **Many of us are often "not there":** Ellen J. Langer, "Mindful Learning," *Current Directions in Psychological Science* 9, no. 6 (December 1, 2000): 220–23, https://doi.org/10.1111/1467-8721.00099.

128 **We rationalize our self-centeredness:** "It's not that we're depressed," Peter Bregman writes, "it's that we're *distracted*. And laughter, it turns out, is not something that happens when we're distracted." Bregman, "Why You Should Treat Laughter as a Metric," *Harvard Business Review*, December 12, 2013, https://hbr.org/2013/12/why-you-should-treat-laughter-as-a-metric.

130 **couples who resist dialed-in responses:** Leslie C. Burpee and Ellen J. Langer, "Mindfulness and Marital Satisfaction," *Journal of Adult Development* 12, no. 1 (January 2005): 43–51, https://doi.org/10.1007/s10804-005-1281-6.

130 **"a relationship stays stable":** Burpee and Langer, "Mindfulness and Marital," 50.

130 **Gallup researchers asked:** "Gallup 2018 Global Emotions Report," Gallup, accessed February 10, 2021, www.gallup.com/analytics/241961/gallup-global-emotions-report-2018.aspx.

130 **Finding humor in everyday life:** Masahiro Toda et al., "Effect of Laughter on Salivary Endocrinological Stress Marker Chromogranin A," *Biomedical Research* 28, no. 2 (April 2007): 115–18, https://doi.org/ 10.2220/biomedres.28.115.

131 **The more humor in your life:** Nicolas A. Kuiper et al., "Coping Humour, Stress, and Cognitive Appraisals," *Canadian Journal of Behavioural Science* 25, no. 1 (January 1993): 81–96, https://doi.org/10.1037/h0078791.

131 **Laughing is primarily:** Robert R. Provine and Kenneth R. Fischer, "Laughing, Smiling, and Talking: Relations to Sleeping and Social Context in Humans," *Ethology* 83 (1989): 295–305, https://psychology.umbc.edu/files/2013/10/Laughing-Smiling-and-Talking-Relation-to-Sleeping-and-social-context-in-humans-Robert-R.Provine.pdf.

131 **asked three hundred people:** Ryan J. Dwyer et al., "Smartphone Use Undermines Enjoyment of Face-to-Face Social Interactions," *Journal of Experimental Social Psychology* 78 (September 2018): 233–39, https://doi.org/10.1016/j.jesp.2017.10.007.

131 **Senior researcher Elizabeth Dunn:** Dwyer et al., "Smartphone Use Undermines," 233–239.

132 **Conversations can seem more superficial:** Shalini Misra et al., "The iPhone Effect: The Quality of In-Person Social Interactions in the Presence of Mobile Devices," *Environment and Behavior* 48, no. 2 (February 1, 2016): 275–98, https://doi.org/10.1177/0013916514539755.

132 **89 percent of smartphone users:** Lee Rainie and Kathryn Zickuhr, "Americans' Views on Mobile Etiquette," Pew Research Center, August 26, 2015, www.pewinternet.org/2015/08/26/americans-views-on-mobile-etiquette/.

132 **compromising the quality:** Hunt Allcott et al., *The Welfare Effects of Social Media* (working paper 25514, National Bureau of Economic Research, Cambridge, MA, November 2019), www.nber.org/papers/w25514. After four weeks of quitting Facebook, participants reported spending more in-person time with friends and family, better daily moods, greater life satisfaction, and an extra hour of downtime a day.

132 **Participants in a Yale University study:** Boothby et al., "Shared Experiences Are Amplified," *Psychological Science* 25, no. 12 (December 1, 2014), 2209–16, https://doi.org/10.1177/0956797614551162.

132 **Researcher Erica Boothby explains:** Association for Psychological Science, "Sharing Makes Both Good and Bad Experiences More Intense," October 7, 2014, www.psychologicalscience.org/news/releases/sharing-makes-both-good-and-bad-experiences-more-intense.html.

132 **Your cell phone may:** Samantha Boardman, "Is Your Cellphone Ruining Your Relationship?," Positive Prescription (blog), accessed January 26, 2021, https://positiveprescription.com/is-your-cell-phone-ruining-your-relationship-2/.

132 **"Partner phone snubbing":** James A. Roberts and Meredith E. David, "My Life Has Become a Major Distraction from My Cell Phone: Partner Phubbing and Relationship Satisfaction among Romantic Partners," *Computers in Human Behavior* 54 (August 2015): 134–41, https://doi.org/10.1016/j.chb.2015.07.058.

133 **People who turn warmly:** John M. Gottman and Julie Gottman, "The Natural Principles of Love," *Journal of Family Theory & Review* 9, no. 1 (March 2, 2017): 7–26, https://doi.org/10.1111/jftr.12182.

Part Four: Challenge Yourself and Embody Vitality

CHAPTER 16: CONSTRUCTIVE NEGATIVITY

138 **British psychiatrist Colin Murray Parkes:** Colin M. Parkes and Holly G. Prigerson, *Bereavement: Studies of Grief in Adult Life*, 4th ed. (New York: Routledge, 2013), 6.

139 **On those occasions:** Allison S. Troy et al., "A Person-by-Situation Approach to Emotion Regulation: Cognitive Reappraisal Can Either Help or Hurt, Depending on the Context," *Psychological Science* 24, no. 12 (October 2013), https://doi.org/10.1177/0956797613496434.

139 **Habitual suppression comes:** Benjamin P. Chapman et al., "Emotion Suppression and Mortality Risk over a 12-Year Follow-Up," *Journal of Psychosomatic Research* 75, no. 4 (October 2013): 381–85, https://doi.org/10.1016/j.jpsychores.2013.07.014.

139 **Suppressors are also:** James J. Gross et al., "Emotion Regulation in Everyday Life," in *Emotion Regulation in Couples and Families: Pathways to Dysfunction and Health*, eds. Douglas K. Snyder, Jeffry A. Simpson, and Jan N. Hughes (Washington, DC: American Psychological Association, 2006), 13–35, https://doi.org/10.1037/11468-001.

141 **Feeling disappointment better enables:** Noelle Nelson et al., "Emotions Know Best: The Advantage of Emotional versus Cognitive Responses to Failure," *Journal of Behavioral Decision Making* 31, no. 2 (September 2017): 40–51, https://doi.org/10.1002/bdm.2042.

143 **Distressing feelings are less likely:** Todd B. Kashdan et al., "Unpacking Emotion Differentiation: Transforming Unpleasant Experience by Perceiving Distinctions in Negativity," *Current Directions in Psychological Science* 24, no. 1 (February 18, 2015): 10–16, https://doi.org/10.1177/0963721414550708.

143 **might manifest later:** Samantha Boardman, "How to Get Out of a Bad Mood" *Marie Claire*, May 28, 2019, https://www.marieclaire.com/health-fitness/a27582877/bad-mood-tips/.

144 **Evidence suggests that people:** Jonathan M. Adler and Hal E. Hershfield, "Mixed Emotional Experience Is Associated with and Precedes Improvements in Psychological Well-Being," *PLoS ONE* 7, no. 4 (April 23, 2012): e35633, https://journals.plos.org/plosone/article?id=10.1371/journal.pone.0035633.

144 **wide mix of emotions—emodiversity:** Jordi Quoidbach et al., "Emodiversity and the Emotional Ecosystem," *Journal of Experimental Psychology: General* 143, no. 6 (October 2014): 2057–66, https://doi.org/10.1037/a0038025.

144 **A bad mood doesn't last:** Gloria Luong et al., "When Bad Moods May Not Be So Bad: Valuing Negative Affect Is Associated with Weakened Affect-Health

Links," *Emotion* 16, no. 3 (September 2016): 387, https://doi.org/10.1037/emo0000132.

144 **Attitude and perception:** Asmir Gračanin et al., "Why Crying Does and Sometimes Does Not Seem to Alleviate Mood: A Quasi-Experimental Study," *Motivation and Emotion* 39, no. 6 (August 23, 2015): https://doi.org/10.1007/s11031-015-9507-9.

144 **Health researchers at Penn State:** Science News, "Let It Go: Reaction to Stress More Important Than Its Frequency," *ScienceDaily*, February 25, 2016, www.sciencedaily.com/releases/2016/02/160225140246.htm.

145 **getting an electric shock:** Giles W. Story et al., "Dread and the Disvalue of Future Pain," *PLoS Computational Biology* 9, no. 11 (November 21, 2013): e1003335, https://doi.org/10.1371/journal.pcbi.1003335.

147 **Psychologist Karen Reivich:** Karen Reivich and Andrew Shatté, *The Resilience Factor: 7 Keys to Finding Your Inner Strength and Overcoming Life's Hurdles* (New York: Harmony, 2003), 123–144.

149 **Most of us have great affection:** James E. Groves, "Taking Care of the Hateful Patient," *New England Journal of Medicine* 298, no. 16 (May 1978): 883–87, https://doi.org/10.1056/NEJM197804202981605. "Admitted or not, the fact remains that a few patients kindle aversion, fear, despair, or even downright malice in their doctors," Dr. Groves wrote in a seminal 1978 paper, "Taking Care of the Hateful Patient," published in the *New England Journal of Medicine*. Groves divided these challenging patients into the following four groups:

Dependent Clingers: Dependent Clingers have a relentless need for attention and reassurance. No matter how much time and energy a doctor gives them, it is never enough. They exhaust the doctor with endless phone calls and questions—which have usually been covered already. These patients lack boundaries and try to reach the doctor at any time, day or night, and disregard on-call protocols. "Whatever their medical problems, what is common to them as a group is their . . . bottomless need."

Entitled Demanders: Entitled Demanders are similar to Dependent Clingers in terms of their neediness but are more overtly and obnoxiously demanding. They become hostile when they do not get what they want. Intimidation and devaluation are their go-to strategies. Typical behaviors include threatening legal action, damaging the doctor's reputation, and withholding payment if the doctor isn't complying with their stipulations.

Manipulative Help-Rejecters: Manipulative Help-Rejecters are the patients who feel that nothing and nobody can help them. As Gross observes: "Appearing almost smugly satisfied, they return again and again to the office or clinic to report that, once again, the regimen did not work. Their pessimism and tenacious nay-saying appear to increase in direct proportion to the physician's efforts and enthusiasm." It's as if they want us to fail.

Self-Destructive Deniers: Self-destructive deniers seem to revel in self-destruction. "They appear to find their main pleasure in furiously defeating the physician's attempts to preserve their lives," writes Gross. They have given up hope and don't seem to care about the havoc their behavior wreaks on themselves and those around them.

Although these four categories of challenging patients oversimplify the situation, it's important to recognize that some patients elicit intense negative feelings—aversion, avoidance, anger, anxiety, inadequacy, fear, indifference, and even loathing—instead of wasting precious energy trying to suppress uncomfortable feelings or allowing them to impact quality of care. Instead, I try to utilize them. For example, a feeling of helplessness and inadequacy toward a help-rejecting patient might be a sign of the patient's intense fear of abandonment. A feeling of aversion toward a clingy patient is a sign it is time to set firm boundaries.

"What the behaviors of such patients teach over time is that it is not how one feels about them that is most important in their care. It is how one behaves toward them," wrote Groves.

151 **Even ruminating can be helpful:** Natalie J. Ciarocco et al., "Some Good News about Rumination: Task-Focused Thinking After Failure Facilitates Performance Improvement," *Journal of Social and Clinical Psychology* 29, no. 10 (January 2011): 1057–73, https://doi.org/10.1521/jscp.2010.29.10.1057.

CHAPTER 17: EXPAND YOURSELF

152 **Surveys of leisure time:** US Department of Labor, Bureau of Labor Statistics, *American Time Use Survey—2019 Results*, June 25, 2020, www.bls.gov/news.release/pdf/atus.pdf and US Department of Labor, Bureau of Labor Statistics, *Economic News Release*, June 25, 2020, table 11B, www.bls.gov/news.release/atus.t11B.htm.

152 **Screens trump socializing:** Chris Weller, "Here's Why It Feels Like You Have No Free Time, in One Chart," *Business Inside*, May 12, 2017, www.businessinsider.com/no-free-time-adam-alter-2017-5.

153 **"The Paradox of Happiness," explores:** L. Parker Schiffer and Tomi-Ann Roberts, "The Paradox of Happiness: Why Are We Not Doing What We Know Makes Us Happy?," *Journal of Positive Psychology* 13, no. 3 (May 4, 2018): 252–59, https://doi.org/10.1080/17439760.2017.1279209.

153 **They are called "demand shielding":** Chen Zhang et al., "More Is Less: Learning But Not Relaxing Buffers Deviance Under Job Stressors," *Journal of Applied Psychology* 103, no. 2 (2018): 123, http://dx.doi.org/10.1037/apl0000264.

153 **Learning something novel:** Chen Zhang et al., "More Is Less," 123.

154 **Astronomy professor Abraham Loeb:** Focal Point Series, "One Thing to Change: Think More Like Children," June 28, 2019, *Harvard Gazette*, https://news.harvard.edu/gazette/story/2019/06/focal-point-harvard-professor-avi-loeb-wants-more-scientists-to-think-like-children/.

154 **During one study, participants were:** Brent A. Mattingly and Gary W. Lewandowski Jr., "The Power of One: Benefits of Individual Self-Expansion," *Journal of Positive Psychology* 8, no. 1 (January 2013): 12–22, https://doi.org/10.1080

/17439760.2012.746999. Self-expansion is typically cited in the context of romantic relationships but also has application for individuals.

154 **To assess daily positive experiences:** "Gallup 2019 Global Emotions Report," Gallup, accessed February 10, 2021, www.gallup.com/analytics/248906/gallup-global-emotions-report-2019.aspx and https://www.amateo.info/Site/Data/1939/Files/Gallup%20Wellbeing%20Index%20-%20AMATEO%20FINAL.pdf.

154 **To determine if an activity:** Mattingly and Lewandowski Jr., "The Power of One," 12–14.

155 **Functional imaging reveals:** Xiaomeng Xu et al., "An fMRI Study of Nicotine-Deprived Smokers' Reactivity to Smoking Cues during Novel/Exciting Activity," *PLoS ONE* 9, no. 4 (April 11, 2014): e94598, https://doi.org/10.1371/journal.pone.0094598.

155 **known as "self-complexity":** Patricia W. Linville, "Self-Complexity as a Cognitive Buffer Against Stress-Related Illness And Depression," *Journal of Personality and Social Psychology* 52, no. 4 (1987): 663, https://doi.org/10.1037//0022-3514.52.4.663.

155 **Not placing all your eggs:** Sarah E. Gaither et al., "Thinking about Multiple Identities Boosts Children's Flexible Thinking," *Developmental Science* 23, no. 2 (May 30, 2019): e12871, https://doi.org/10.1111/desc.12871.

156 **Flow experiences facilitate well-being:** Angela Lee Duckworth et al., "Positive Psychology in Clinical Practice," *Annual Review of Clinical Psychology* 1, no. 1 (February 2005): 629–51, https://doi.org/10.1146/annurev.clinpsy.1.102803.144154.

157 **Albert Einstein's letter:** Maria Popova, "The Secret to Learning Anything: Albert Einstein's Advice to His Son," BrainPickings, accessed January 29, 2021, www.brainpickings.org/2013/06/14/einstein-letter-to-son/.

157 **Flow isn't limited:** Josh Jones, "Albert Einstein Tells His Son the Key to Learning and Happiness Is Losing Yourself in Creativity (or 'Finding Flow')," OpenCulture, May 26, 2015 www.openculture.com/2015/05/einstein-tells-his-son-the-key-to-learning-happiness-is-losing-yourself-in-creativity.html. This piece gives a nice explanation of flow.

157 **The more flow you experience:** Wladislaw Rivkin et al., "Which Daily Experiences Can Foster Well-Being at Work? A Diary Study on the Interplay Between Flow Experiences, Affective Commitment, and Self-Control Demands," *Journal of Occupational Health Psychology* 23, no. 1 (January 2018): 99, https://doi.org/10.1037/ocp0000039.

158 **Putting flow back into everyday life:** Mihály Csíkszentmihályi, "Flow, the Secret to Happiness," filmed in 2004, TED video, 18:41, www.ted.com/talks/mihaly_csikszentmihalyi_on_flow/transcript?language=en. This is the transcript for Csíkzentmihályi's TED talk that explains it so well.

159 **Making art reduces cortisol:** Girija Kaimal et al., "Reduction of Cortisol Levels and Participants' Responses Following Art Making," *Journal of the American Art Therapy Association* 33, no. 2 (May 23, 2016): 74, https://doi.org/10.1080/07421656.2016.1166832.

160 **For people who are obsessed:** Kevin J. Eschleman et al., "Benefiting from Creative Activity: The Positive Relationships between Creative Activity, Recovery Experiences, and Performance-Related Outcomes," *Journal of Occupational and Organizational Psychology* 87, no. 3 (April 17, 2014): 579–98, https://doi.org/10.1111/joop.12064.

160 **Hobbies create a sense:** Eschleman et al., "Benefiting from Creative," 583–86.

160 **"Many scientists say":** Julia Rosen, "How a Hobby Can Boost Researchers' Productivity and Creativity," *Nature*, June 18, 2018, www.nature.com/articles/d41586-018-05449-7.

160 **Nobel Prize winners:** Robert Root-Bernstein et al., "Arts Foster Scientific Success: Avocations of Nobel, National Academy, Royal Society, and Sigma Xi Members," *Journal of Psychology of Science and Technology* 1, no. 2 (October 2008): 51–63, https://doi.org/10.189/1939-7054.1.2.51.

161 **Physicists and professional writers reported:** Shelly L. Gable et al., "When the Muses Strike: Creative Ideas of Physicists and Writers Routinely Occur During Mind Wandering," *Psychological Science* 30, no. 3 (March 2019): 396–404, https://doi.org/10.1177/0956797618820626.

161 **"best thing for being sad":** T. H. White, *The Once and Future King*, reprint ed. (New York: Ace Books, 1987), 183.

CHAPTER 18: EMBODIED HEALTH

162 **A healthy appearance:** Brian R. Spisak et al., "A Face for All Seasons: Searching for Context-Specific Leadership Traits and Discovering a General Preference for Perceived Health," *Frontiers in Human Neuroscience* 8 (November 2014): 792, https://doi.org/10.3389/fnhum.2014.00792.

162 **Just one night of poor sleep:** Nancy L. Sin et al., "Bidirectional, Temporal Associations of Sleep with Positive Events, Affect, and Stressors in Daily Life across a Week," *Annals of Behavioral Medicine* 51, no. 3 (February 10, 2017): 402–15, https://doi.org/10.1007/s12160-016-9864-y.

162 **Not getting enough rest:** Ninad Gujar et al., "A Role For REM Sleep in Recalibrating the Sensitivity of the Human Brain to Specific Emotions," *Cerebral Cortex* 21, no. 1 (April 26, 2010): 115–23, https://doi.org/10.1093/cercor/bhq064.

162 **more vulnerable to "cognitive interference":** Soomi Lee et al., "Bidirectional Associations of Sleep with Cognitive Interference in Employees' Work Days," *Sleep Health* 5, no. 3 (June 2019): 298–308, https://doi.org/10.1016/j.sleh.2019.01.007.

163 **High school students who slept:** Matthew D. Weaver et al., "Dose-Dependent Associations between Sleep Duration and Unsafe Behaviors among US High School Students," *JAMA Pediatrics* 172, no. 12 (October 1, 2018): 1187–89, https://doi.org/https://doi.org/10.1001/jamapediatrics.2018.2777.

163 **A UC Berkeley study showed:** Eti Ben Simon and Matthew P. Walker, "Sleep Loss Causes Social Withdrawal and Loneliness," *Nature Communications* 9, no. 3146 (August 14, 2018): 1, https://doi.org/10.1038/s41467-018-05377-0.

163 **"perceive you as 'socially repulsive'":** Yasmin Anwar, "Poor Sleep Triggers Viral Loneliness and Social Rejection," *Berkeley News*, August 14, 2018, https://news.berkeley.edu/2018/08/14/sleep-viral-loneliness/.

163 **Interacting with a rude:** Charlotte Fritz et al., "Workplace Incivility Ruins My Sleep and Yours: The Costs of Being in a Work-Linked Relationship," *Occupational Health Science* 3 (December 11, 2018): 1–21, https://doi.org/10.1007/s41542-018-0030-8.

163 **the world's largest sleep study:** Conor J. Wild et al., "Dissociable Effects of Self-Reported Daily Sleep Duration on High-Level Cognitive Abilities," *Sleep* 41, no. 12 (December 1, 2018): 1–11, https://doi.org/10.1093/sleep/zsy182.

163 **Americans are sleeping six hours:** Connor M. Sheehan et al., "Are U.S. Adults Reporting Less Sleep?: Findings from Sleep Duration Trends in the National Health Interview Survey, 2004–2017," *Sleep* 42, no. 2 (February 2019): 1–8, https://doi.org/10.1093/sleep/zsy221.

163 **We interpret neutral information:** Daniela Tempesta et al., "The Impact of Five Nights of Sleep Restriction on Emotional Reactivity," *Journal of Sleep Research* (October 2020): e13022, https://doi.org/10.1111/jsr.13022.

164 **Most people recognize that sleep:** *2018 Sleep Prioritization and Personal Effectiveness*, SleepFoundation, https://www.sleepfoundation.org/wp-content/uploads/2018/10/Sleep-in-America-2018_prioritizing-sleep.pdf.

164 **You may sleep so long:** Samantha Boardman, "Set Your Alarm to GO to Bed," *The Wall Street Journal*, April 14, 2016, https://www.wsj.com/articles/BL-258B-6919.

164 **Sixty-eight percent of phone owners:** Matthew A. Christensen et al., "Direct Measurements of Smartphone Screen-Time: Relationships with Demographics and Sleep," *PLoS ONE* 11, no. 11 (November 9, 2016): e0165331, https://doi.org/10.1371/journal.pone.0165331.

165 **We think we're fine:** Hans P. A. Van Dongen et al., "The Cumulative Cost of Additional Wakefulness: Dose-Response Effects on Neurobehavioral Functions and Sleep Physiology from Chronic Sleep Restriction and Total Sleep Deprivation," *Sleep* 26, no. 2 (March 2003): 117–26, https://doi.org/10.1093/sleep/26.2.117.

165 **It is a virtuous cycle:** Soomi Lee et al., "Daily Antecedents and Consequences of Nightly Sleep," *Journal of Sleep Research* 26, no. 4 (December 23, 2016): 498–509, https://doi.org/doi:10.1111/jsr.12488.

165 **Professor Orfeu Buxton:** Marjorie S. Miller, "Good Sleep May Promote Positive Experiences, Less Conflict," *Penn State News*, February 15, 2017, https://news.psu.edu/story/450884/2017/02/15/research/good-sleep-may-promote-positive-experiences-less-conflict.

166 **Today, it's been shown:** James A. Blumenthal et al., "Is Exercise a Viable Treatment for Depression?," *American College of Sports Medicine Health & Fitness Journal* 16, no. 4 (August 2012): 14, https://doi.org/10.1249/01.FIT.0000416000.09526.eb.

166 **Stressed-out university students:** Megan Oaten and Ken Cheng, "Longitudinal Gains In Self-Regulation from Regular Physical Exercise," *British Journal of*

Health Psychology 11, no. 4 (November 2006): 717–33, https://doi.org/10.1348/135910706X96481.

166 **His approach, like my training:** Brett J. Deacon, "The Biomedical Model of Mental Disorder: A Critical Analysis of Its Validity, Utility, and Effects on Psychotherapy Research," *Clinical Psychology Review* 33, no. 7 (April 2013): 846–61, https://doi.org/10.1016/j.cpr.2012.09.007.

167 **Four out of five patients:** Carol A. Janney et al., "Moving Towards Wellness: Physical Activity Practices, Perspectives, and Preferences of Users of Outpatient Mental Health Service," *General Hospital Psychiatry* 49 (November 2017): 63–66, https://doi.org/10.1016/j.genhosppsych.2017.07.004.

167 **a sedentary lifestyle almost doubled:** Samuel B. Harvey et al., "Exercise and the Prevention of Depression: Results of the HUNT Cohort Study," *American Journal of Psychiatry* 175, no. 1 (January 1, 2018): 28–36, https://doi.org/10.1176/appi.ajp.2017.16111223 and Karmel W. Choi et al., "Assessment of Bidirectional Relationships between Physical Activity and Depression among Adults: A 2-Sample Mendelian Randomization Study," *Journal of American Medical Association Psychiatry* 76, no. 4 (April 1, 2019): 399–408, https://doi.org/10.1001/jamapsychiatry.2018.4175.

167 **more than a million people:** Sammi R. Chekroud et al., "Association between Physical Exercise and Mental Health in 1·2 Million Individuals in the USA between 2011 and 2015: A Cross-Sectional Study, *Lancet Psychiatry* 5, no. 9 (September 2018): 739–46, https://doi.org/10.1016/S2215-0366(18)30227-X.

168 **regular, everyday actions:** Neal Lathia et al., "Happier People Live More Active Lives: Using Smartphones to Link Happiness and Physical Activity," *PLoS ONE* 12, no. 1 (January 4, 2017): e0160589, https://doi.org/10.1371/journal.pone.0160589.

168 **Comprehensive gait analysis:** Johannes Michalak et al., "How We Walk Affects What We Remember: Gait Modifications through Biofeedback Change Negative Affective Memory Bias," *Journal of Behavior Therapy and Experimental Psychiatry* 46 (March 2015): 121, https://doi.org/10.1016/j.jbtep.2014.09.004.

168 **Without even looking:** Hillel Aviezer et al., "Body Cues, Not Facial Expressions, Discriminate between Intense Positive and Negative Emotions," *Science* 338, no. 6111 (November 30, 2012): 1225, https://doi.org/10.1126/science.1224313.

168 **who were prompted to imitate:** Michalak et al., "How We Walk," 121.

169 **Good posture can even bolster:** Erik Peper et al., "Do Better in Math: How Your Body Posture May Change Stereotype Threat Response," *NeuroRegulation* 5, no. 2 (June 29, 2018): 67, https://doi.org/10.15540/nr.5.2.67.

169 **a supine position reduces defensiveness:** Tom F. Price and Eddie Harmon-Jones, "Embodied Emotion: The Influence of Manipulated Facial and Bodily States on Emotive Responses," *Wiley Interdisciplinary Reviews: Cognitive Science* 6, no. 6 (September 24, 2015): 461–73, https://doi.org/10.1002/wcs.1370.

170 **"mind if I go get one?":** Katie Heaney, "The Single Easiest Way to Control Your Emotions at Work," *The Cut*, June 7, 2018, https://www.thecut.com/2018/06/how-to-control-your-emotions-at-work.html?.

171 **Given the finding:** "How Much Time Do We Spend On Social Media?," MediaKix, accessed January 29, 2021, http://mediakix.com/how-much-time-is-spent-on-social-media-lifetime/#gs.bz07wq.

171 **"Often when we feel depleted":** Science News, "Spending Time in Nature Makes People Feel More Alive, Study Shows," *ScienceDaily*, June 4, 2010, www.sciencedaily.com/releases/2010/06/100603172219.htm.

171 **Just twenty minutes in nature:** MaryCarol R. Hunter et al., "Urban Nature Experiences Reduce Stress in the Context of Daily Life Based on Salivary Biomarkers," *Frontiers in Psychology* 10 (April 4, 2019): 722, https://doi.org/10.3389/fpsyg.2019.00722.

171 **twelve thousand-plus Americans:** "A National Initiative to Understand and Connect Americans and Nature," Nature of Americans, accessed January 29, 2021, https://natureofamericans.org/findings.

171 **A ninety-minute walk in green space:** Gregory N. Bratman et al., "Nature Experience Reduces Rumination and Subgenual Prefrontal Cortex Activation," *Proceedings of the National Academy of Sciences* 112, no. 28 (July 2015): 8567–72, https://doi.org/10.1073/pnas.1510459112.

172 **being in nature makes us:** Nicolas Guéguen and Jordy Stefan, "'Green Altruism': Short Immersion in Natural Green Environments and Helping Behavior," *Environment and Behavior* 48, no. 2 (February 1, 2016): 324–42, https://doi.org/10.1177/0013916514536576.

172 **When mothers and daughters walk:** Dina Izenstark and Aaron T. Ebata, "The Effects of the Natural Environment on Attention and Family Cohesion: An Experimental Study," *Children, Youth and Environments* 27, no. 2 (January 2017): 93, https://doi.org/10.7721/chilyoutenvi.27.2.0093.

CHAPTER 19: FORTIFY THE BODY, FORTIFY THE MIND

173 **researchers conducting a 2015 British study:** Tamlin S. Conner et al., "On Carrots and Curiosity: Eating Fruit and Vegetables Is Associated with Greater Flourishing in Daily Life," *British Journal of Health Psychology* 20, no. 2 (April 3, 2015): 413–27, https://doi.org/10.1111/bjhp.12113. "These findings suggest that FV (fruit and vegetable) intake is related to . . . aspects of human flourishing, beyond just feeling happy. These results are important because eudaemonic well-being is thought to play an important role in psychological resilience (Steger, Kashdan, & Oishi, 2008). It is worth noting that the Curiosity has also been linked to greater resilience (Kashdan, 2009)."

173 **A person's ability to focus:** Annelise A. Madison et al., "Afternoon Distraction: A High-Saturated-Fat Meal and Endotoxemia Impact Postmeal Attention in a Randomized Crossover Trial," *American Journal of Clinical Nutrition* 111, no. 6 (June 1, 2020): 1150–58, https://doi.org/10.1093/ajcn/nqaa085.

173 **Four days of eating:** Tuki Attuquayefio et al., "A Four-Day Western-Style Dietary Intervention Causes Reductions in Hippocampal-Dependent Learning and Memory and Interoceptive Sensitivity," *PLoS ONE* 12, no. 2 (February 23,

2017): e0172645, https://doi.org/10.1371/journal.pone.0172645 and Cameron J. Holloway et al., "A High-Fat Diet Impairs Cardiac High-Energy Phosphate Metabolism and Cognitive Function in Healthy Human Subjects," *American Journal of Clinical Nutrition* 93, no. 4 (April 2011): 748–55, https://doi.org/10.3945/ajcn.110.002758. From the latter study: "Subjects receiving the high-fat diet had abnormalities in cognition, which was measured by using a validated computerized test battery (29). Deficits were found in the speed of retrieval of information from memory, the ability to intensely focus attention, and performance of a complex higher order task involving working memory and attention. Perhaps not surprisingly, contentedness was reduced in those receiving the high-fat diet."

174 **On the flip side, researchers:** Heather M. Francis et al., "A Brief Diet Intervention Can Reduce Symptoms of Depression in Young Adults—A Randomised Controlled Trial," *PLoS ONE* 14, no. 10 (October 2019): e0222768, https://doi.org/10.1371/journal.pone.0222768.

174 **Nutrition has mostly been:** Jerome Sarris et al., "Nutritional Medicine as Mainstream in Psychiatry," *Lancet Psychiatry* 2, no. 3 (March 2015): 271–74, https://doi.org/10.1016/S2215-0366(14)00051-0.

174 **If one of them orders:** Eric Robinson et al., "Social Influences on Eating: Implications for Nutritional Interventions," *Nutrition Research Reviews* 26, no. 2 (October 8, 2013): 166–76, https://doi.org/10.1017/S0954422413000127.

174 **you may feel tempted:** Suzanne Higgs and Jason Thomas, "Social Influences on Eating," *Current Opinion in Behavioral Sciences* 9 (June 2016): 1–6, https://doi.org/10.1016/j.cobeha.2015.10.005.

174 **Eating with friends often leads:** John M. de Castro, "Family and Friends Produce Greater Social Facilitation of Food Intake Than Other Companions," *Physiology & Behavior* 56, no. 3 (September 1994): 445–55, https://doi.org/10.1016/0031-9384(94)90286-0.

174 **Instead of relying on internal signals:** Lenny R. Vartanian et al., "Modeling of Food Intake: A Meta-Analytic Review," *Social Influence* 10, no. 3 (July 3, 2015): 119–36, https://doi.org/10.1080/15534510.2015.1008037.

175 **Placing dishes at the top:** Eran Dayan and Maya Bar-Hillel, "Nudge to Nobesity II: Menu Positions Influence Food Orders," *Judgment and Decision Making* 6, no. 4 (June 2011): 33342.

175 **Elaborate descriptions:** Nicolas Guéguen and Céline Jacob, "The Effect of Menu Labels Associated with Affect, Tradition and Patriotism on Sales," *Food Quality and Preference* 23, no. 1 (January 2012): 86–88, https://doi.org/10.1016/j.foodqual.2011.07.001.

175 **Size also matters:** Nicole Diliberti et al., "Increased Portion Size Leads to Increased Energy Intake in a Restaurant Meal," *Obesity Research* 12, no. 3 (September 6, 2012): 562–68, https://doi.org/10.1038/oby.2004.64.

175 **We tend to consume more:** Richard Hébert, "The Weight Is Over," Association for Psychological Science, January 19, 2005, www.psychologicalscience.org/observer/the-weight-is-over.

175 **Workday stress can also lead:** Yihao Liu et al., "Eating Your Feelings? Testing a Model of Employees' Work-Related Stressors, Sleep Quality, and Unhealthy Eating," *Journal of Applied Psychology* 102, no. 8 (April 2017): 1237, https://doi.org/10.1037/apl0000209.

175 **Feeling physically exhausted:** Wen Lv et al., "Sleep, Food Cravings and Taste," *Appetite* 125 (June 1, 2018): 210–16, https://doi.org/10.1016/j.appet.2018.02.013.

175 **Levels of the hunger hormone ghrelin:** Lisa M. Jaremka et al., "Novel Links between Troubled Marriages and Appetite Regulation: Marital Distress, Ghrelin, and Diet Quality," *Clinical Psychological Science* 4, no. 3 (May 2016): 363–75, https://doi.org/10.1177/2167702615593714.

175 **quality of their daily encounters:** Kharah Ross et al., "Social Encounters in Daily Life and 2-Year Changes in Metabolic Risk Factors in Young Women," *Development and Psychopathology* 23, no. 3 (August 2011): 897–906, https://doi.org/10.1017/S0954579411000381.

176 **The closer we eat to bedtime:** Cibele Aparecida Crispim et al., "Relationship between Food Intake and Sleep Pattern in Healthy Individuals," *Journal of Clinical Sleep Medicine* 7, no. 6 (December 15, 2011): 659–64, https://doi.org/10.5664/jcsm.1476.

176 **risk of becoming "hangry":** Jennifer K. MacCormack and Kristen A. Lindquist, "Feeling Hangry? When Hunger Is Conceptualized as Emotion," *Emotion* 19, no. 2 (March 2019): 301–19, https://doi.org/10.1037/emo0000422.

176 **Researchers analyzed parole hearings:** Shai Danziger et al., "Extraneous Factors in Judicial Decisions," *Proceedings of the National Academy of Sciences* 108, no. 17 (April 26, 2011): 6889–92, https://doi.org/10.1073/pnas.1018033108.

176 **Pay attention to the signals:** "Are You Really You When You're Hungry?," American Psychological Association, June 11, 2018, www.apa.org/news/press/releases/2018/06/hungry.

177 **simultaneous changes often reinforce:** Michael D. Mrazek et al., "Pushing the Limits: Cognitive, Affective, and Neural Plasticity Revealed by an Intensive Multifaceted Intervention," *Frontiers in Human Neuroscience* 10 (March 18, 2016): 117, https://doi.org/10.3389/fnhum.2016.00117.

Part Five: Beyond You

CHAPTER 20: CONTRIBUTE VALUE

182 **Putting yourself on a pedestal:** S. Katherine Nelson-Coffey et al., "Do Unto Others or Treat Yourself? The Effects of Prosocial and Self-Focused Behavior on Psychological Flourishing," *Emotion* 16, no. 6 (December 2016): 850, https://doi.org/10.1037/emo0000178.

183 **People also tend to feel better:** Elizabeth W. Dunn et al., "Spending Money on Others Promotes Happiness," *Science* 319, no. 5870 (March 21, 2008): 1687–88, https://doi.org/10.1126/science.1150952.

183 ***Having* can get boring:** In chapter 8 I pointed out that people tend to adapt to nice things. It doesn't take long to get used to a source of pleasure if we experience it over and over again. Hedonic adaptation explains why the joy you feel

upon getting something new doesn't last. But the feeling of fulfillment we get from giving resists the corrosive decay of hedonic adaptation.

183 **Margot was dwelling on how:** Sara B. Algoe et al., "Putting the 'You' in 'Thank You': Examining Other-Praising Behavior as the Active Relational Ingredient in Expressed Gratitude," *Social Psychological and Personality Science* 7, no. 7 (September 2016): 658–66, https://doi.org/10.1177/1948550616651681.

183 **Feeling "in the loop":** Heidi Grant, "Stop Making Gratitude All about You," *Harvard Business Review*, June 29, 2016, https://hbr.org/2016/06/stop-making-gratitude-all-about-you.

184 **I explained to Margot:** Iris B. Mauss et al., "The Pursuit of Happiness Can Be Lonely," *Emotion* 12, no. 5 (October 2012): 908, https://doi.org/10.1037/a0025299.

184 **"In modern society":** John T. Cacioppo et al., "Reciprocal Influences between Loneliness and Self-Centeredness: A Cross-Lagged Panel Analysis in a Population-Based Sample of African American, Hispanic, and Caucasian Adults," *Personality and Social Psychology Bulletin* 43, no. 8 (August 2017): 1125–35, https://doi.org/10.1177/0146167217705120. Cacioppo's quote was sourced from Science News, "Loneliness Contributes to Self-Centeredness for Sake of Self-Preservation," *ScienceDaily*, June 13, 2017, www.sciencedaily.com/releases/2017/06/170613102013.htm.

185 **Paying too much attention:** Nilly Mor and Jennifer Winquist, "Self-Focused Attention and Negative Affect: A Meta-Analysis," *Psychological Bulletin* 128, no. 4 (August 2002): 638, https://doi.org/10.1037/0033-2909.128.4.638.

185 **Being kind and generous:** Thaddeus W. W. Pace et al., "Effect of Compassion Meditation on Neuroendocrine, Innate Immune and Behavioral Responses to Psychosocial Stress," *Psychoneuroendocrinology* 34, no. 1 (January 2009): 87–98, https://doi.org/10.1016/j.psyneuen.2008.08.011.

185 **lessen symptoms of depression:** Kristin Layous et al., "Delivering Happiness: Translating Positive Psychology Intervention Research for Treating Major and Minor Depressive Disorders," *Journal of Alternative and Complementary Medicine* 17, no. 8 (August 2011): 675–83, https://doi.org/10.1089/acm.2011.0139.

185 **having an empathetic doctor:** Kari A. Leibowitz et al., "Physician Assurance Reduces Patient Symptoms in US Adults: An Experimental Study," *Journal of General Internal Medicine* 33, no. 12 (August 20, 2018): 2051–52, https://doi.org/10.1007/s11606-018-4627-z.

186 **Nearly 50 percent of Americans:** Frank Newport, "Americans' Perceived Time Crunch No Worse Than in Past," Gallup, December 31, 2015, https://news.gallup.com/poll/187982/americans-perceived-time-crunch-no-worse-past.aspx.

186 **described as "time famine":** Leslie A. Perlow, "The Time Famine: Toward a Sociology of Work Time," *Administrative Science Quarterly* 44, no. 1 (March 1, 1999): 57–81, https://doi.org/10.2307/2667031.

186 **It may be counterintuitive:** Cassie Mogilner et al., "Giving Time Gives You Time," *Psychological Science* 23, no. 10 (October 2012): 1233–38, https://doi.org/10.1177/0956797612442551.

186 **Or as Gwyneth Paltrow:** Marisa Meltzer, "Gwyneth Paltrow Has the Last Laugh," *Town & Country*, April 8, 2020, www.townandcountrymag.com/society/a31942473/gwyneth-paltrow-may-2020-cover-interview-goop/.

187 **In an admittedly odd study:** Andrew R. Todd et al., "Anxious and Egocentric: How Specific Emotions Influence Perspective Taking," *Journal of Experimental Psychology: General* 144, no. 2 (April 2015): 374, https://doi.org/10.1037/xge0000048.

188 **conducted by two Princeton psychologists:** John M. Darley and C. Daniel Batson, "'From Jerusalem to Jericho': A study of Situational and Dispositional Variables in Helping Behavior," *Journal of Personality and Social Psychology* 27, no. 1 (1973): 100, https://doi.org/10.1037/h0034449.

188 **One reason spirituality is thought:** Brick Johnstone et al., "Relationships among Spirituality, Religious Practices, Personality Factors, and Health for Five Different Faith Traditions," *Journal of Religion and Health* 51, no. 4 (December 2012): 1017–41, https://doi.org/10.1007/s10943-012-9615-8.

189 **People who frequently use:** Allison M. Tackman et al., "Depression, Negative Emotionality, and Self-Referential Language: A Multi-Lab, Multi-Measure, and Multi-Language-Task Research Synthesis," *Journal of Personality and Social Psychology* 116, no. 5 (May 2019): 817, https://doi.org/10.1037/pspp0000187, and Science News, "Frequent 'I-Talk' May Signal Proneness to Emotional Distress," *ScienceDaily*, March 6, 2018, www.sciencedaily.com/releases/2018/03/180306115716.htm. University of Arizona researcher Matthias Mehl found that "the average person speaks about 16,000 words a day, about 1,400 of which are, on average, first-person singular pronouns. Those prone to distress may say [those words] up to 2,000 times a day."

CHAPTER 21: MAKE YOURSELF USEFUL

192 **an experimenter dropped a pen:** Felix Warneken and Michael Tomasello, "Altruistic Helping in Human Infants and Young Chimpanzees," *Science* 311, no. 5765 (March 2006): 1301–3, https://doi.org/10.1126/science.1121448.

192 **two-year-old children find helping:** Robert Hepach et al., "The Fulfillment of Others' Needs Elevates Children's Body Posture," *Developmental Psychology* 53, no. 1 (January 2017): 100, https://doi.org/10.1037/dev0000173.

192 **Adding value fulfills:** Edward L. Deci and Richard M. Ryan, "Self-Determination Theory: A Macrotheory of Human Motivation, Development, and Health," *Canadian Psychology/Psychologie Canadienne* 49, no. 3 (August 2008): 182, https://doi.org/10.1037/a0012801. According to self-determination theory (discussed in chapter 3), well-being rests on three key pillars: (1) autonomy (a sense of control), (2) competence (promotes a sense of effectiveness), and (3) relatedness (enhances connection to others).

192 **Experience Corps, a program:** Editorial, "Researchers Find Sustained Improvement in Health in Experience Corps Tutors Over 55," TheSource, March

12, 2009, https://source.wustl.edu/2009/03/researchers-find-sustained-improvement-in-health-in-experience-corps-tutors-over-55/.

193 **When asked about why:** David S. Yeager et al., "Boring but Important: A Self-Transcendent Purpose for Learning Fosters Academic Self-Regulation," *Journal of Personality and Social Psychology* 107, no. 4 (October 2014): 559, https://doi.org/10.1037/a0037637.

194 **Shifting focus away:** Yoona Kang et al., "Effects of Self-Transcendence on Neural Responses to Persuasive Messages and Health Behavior Change," *Proceedings of the National Academy of Sciences* 115, no. 40 (October 2, 2018): 9974–79, https://doi.org/10.1073/pnas.1805573115.

195 **Adam Grant of Wharton:** Adam Grant and David A. Hofmann, "It's Not All about Me: Motivating Hand Hygiene among Health Care Professionals by Focusing on Patients," *Psychological Science* 22, no. 12 (December 2011): 1494–99, https://doi.org/10.1177/0956797611419172.

196 **Rewards are another option:** Edward L. Deci et al., "A Meta-Analytic Review of Experiments Examining the Effects of Extrinsic Rewards on Intrinsic Motivation," *Psychological Bulletin* 125, no. 6 (December 1999): 627, https://doi.org/10.1037/0033-2909.125.6.627.

196 **Research offers a counterintuitive solution:** Lauren Eskreis-Winkler et al., "Dear Abby: Should I Give Advice or Receive It?," *Psychological Science* 29, no. 11 (November 2018): 1797–1806, https://doi.org/10.1177/0956797618795472.

CHAPTER 22: DELIBERATE VITALITY

199 **three groups were given:** Kurt Gray, "The Power of Good Intentions: Perceived Benevolence Soothes Pain, Increases Pleasure, and Improves Taste," *Social Psychological and Personality Science* 3, no. 5 (September 2012): 639–45, https://doi.org/10.1177/1948550611433470.

200 **Setting goals that are specific:** The following are what is known as SMART goals:

Specific: Specific goals work best. A specific goal leaves less room for excuses or losing sight of a goal that may be inconveniently ambiguous. "Lose five pounds" is more effective than "lose weight." "Increase sales by 15 percent," is more effective than "increase sales."

Measurable: Incremental progress keeps us on track. If trying to lose weight, weigh yourself every day and write it down in a journal. If aiming for increased revenue by year's end, track your progress with specific markers. Be data driven every day.

Align with Values: Goals should be consistent with the big picture, whether personal or professional. Avoid shortcuts and remind yourself every day of why you do what you do.

Realistic: By all means, shoot for the stars, but at the same time be sure your goal is within reason, within reach, and within the realm of possibility. Also, a belief in one's ability to succeed is critical for goal-setting; it matters when selecting the degree of difficulty of goals chosen, for sustaining commitment to goals, and for remaining positive even in the face of negative feedback.

Time-Based: Like selecting a measurable goal, a time frame matters for accountability. A stopwatch is a great motivator. Goal-setting not only gets us motivated for a specific target weight, project, or performance but also better enables us to clearly define what to focus on and how to use our attention bandwidth most effectively.

202 **This relentless appetite:** Brian Wansink et al., "Bottomless Bowls: Why Visual Cues of Portion Size May Influence Intake," *Obesity Research* 13, no. 1 (January 2005): 93–100, https://doi.org/10.1038/oby.2005.12.

203 **Poet Ross Gay's:** Ross Gay, *The Book of Delights* (Chapel Hill, NC: Algonquin Books, 2019).

INDEX